ON GUERRILLA WARFARE

On Guerrilla Warfare

MAO TSE-TUNG

Translated from the Chinese

and with an Introduction by Samuel B. Griffith II

UNIVERSITY OF ILLINOIS PRESS

URBANA AND CHICAGO

First Illinois paperback, 2000
© 1961 by Samuel B. Griffith II
Reprinted by arrangement with Jane Griffith
and Belle Gordon Griffith Heneberger

∞ This book is printed on acid-free paper.

Library of Congress Cataloging-in-Publication Data
Mao Tse-tung, 1893–1976.
[Yu chi chan. English]
On guerrilla warfare / Mao Tse-tung ; translated from the Chinese and
with an introduction by Samuel B. Griffith II.
p.   cm.
ISBN 0-252-06892-0 (pbk. : alk. paper)
ISBN 978-0-252-06892-8 (pbk. : alk. paper)
1. Guerrilla warfare.
I. Title.
U240.M343   2000
355'.0218—dc21   00-055194

University of Illinois Press
1325 South Oak Street
Champaign, IL 61820-6903
www.press.uillinois.edu

CONTENTS

INTRODUCTION

I

THE NATURE OF
REVOLUTIONARY GUERRILLA WAR

> . . . the guerrilla campaigns being waged in China
> today are a page in history that has no precedent.
> Their influence will be confined not solely to China
> in her present anti-Japanese struggle, but will be
> world-wide.
>
> —MAO TSE-TUNG, *Yu Chi Chan*, 1937

AT ONE END OF THE SPECTRUM, ranks of electronic boxes buried deep in the earth hungrily consume data and spew out endless tapes. Scientists and engineers confer in air-conditioned offices; missiles are checked by intense men who move about them silently, almost reverently. In forty minutes, countdown begins.

At the other end of this spectrum, a tired man wearing a greasy felt hat, a tattered shirt, and soiled shorts is seated, his back against a tree. Barrel pressed between his knees, butt resting on the moist earth between sandaled feet, is a Browning automatic rifle. Hooked to his belt, two dirty canvas sacks—one holding three home-made bombs, the other four magazines loaded with .30-caliber ammunition. Draped around his neck, a sausage-like cloth tube with

3

three days' supply of rice. The man stands, raises a water bottle to his lips, rinses his mouth, spits out the water. He looks about him carefully, corks the bottle, slaps the stock of the Browning three times, pauses, slaps it again twice, and disappears silently into the shadows. In forty minutes, his group of fifteen men will occupy a previously prepared ambush.

It is probable that guerrilla war, nationalist and revolutionary in nature, will flare up in one or more of half a dozen countries during the next few years. These outbreaks may not initially be inspired, organized, or led by local Communists; indeed, it is probable that they will not be. But they will receive the moral support and vocal encouragement of international Communism, and where circumstances permit, expert advice and material assistance as well.

As early as November, 1949, we had this assurance from China's Number Two Communist, Liu Shao-ch'i, when, speaking before the Australasian Trade Unions Conference in Peking, he prophesied that there would be other Asian revolutions that would follow the Chinese pattern. We paid no attention to this warning.

In December, 1960, delegates of eighty-one Communist and Workers' Parties resolved that the tempo of "wars of liberation" should be stepped up. A month later (January 6, 1961), the Soviet Premier, an unimpeachable authority on "national liberation wars," propounded an interesting series of questions to which he provided equally interesting answers:

Is there a likelihood of such wars recurring? Yes, there is. Are uprisings of this kind likely to recur? Yes, they are. But wars of this kind are popular uprisings. Is there the likelihood of conditions in other countries reaching the point where the cup of the popular patience overflows and they take to arms? Yes, there is such a likelihood. What is the attitude of the Marxists to such uprisings? A most favorable attitude. . . . These uprisings are directed against the corrupt reactionary regimes, against the colonialists. The Communists support just wars of this kind wholeheartedly and without reservations.*

Implicit is the further assurance that any popular movement infiltrated and captured by the Communists will develop an anti-Western character definitely tinged, in our own hemisphere at least, with a distinctive anti-American coloration.

This should not surprise us if we remember that several hundred millions less fortunate than we have arrived, perhaps reluctantly, at the conclusion that the Western peoples are dedicated to the perpetuation of the political, social, and economic *status quo*. In the not too distant past, many of these millions looked hopefully to America, Britain, or France for help in the realization of their justifiable aspirations. But today many of them feel that these aims can be achieved only by a desperate revolutionary struggle that we will probably oppose. This is not a hypothesis; it is fact.

A potential revolutionary situation exists in any country where the government consistently fails in its obligation to ensure at least a minimally decent standard of life for the

* *World Marxist Review*, January, 1961.

great majority of its citizens. If there also exists even the nucleus of a revolutionary party able to supply doctrine and organization, only one ingredient is needed: the instrument for violent revolutionary action.

In many countries, there are but two classes, the rich and the miserably poor. In these countries, the relatively small middle class—merchants, bankers, doctors, lawyers, engineers—lacks forceful leadership, is fragmented by unceasing factional quarrels, and is politically ineffective. Its program, which usually posits a socialized society and some form of liberal parliamentary democracy, is anathema to the exclusive and tightly knit possessing minority. It is also rejected by the frustrated intellectual youth, who move irrevocably toward violent revolution. To the illiterate and destitute, it represents a package of promises that experience tells them will never be fulfilled.

People who live at subsistence level want first things to be put first. They are not particularly interested in freedom of religion, freedom of the press, free enterprise as we understand it, or the secret ballot. Their needs are more basic: land, tools, fertilizers, something better than rags for their children, houses to replace their shacks, freedom from police oppression, medical attention, primary schools. Those who have known only poverty have begun to wonder why they should continue to wait passively for improvements. They see—and not always through Red-tinted glasses—examples of peoples who have changed the structure of their societies, and they ask, "What have we to lose?" When a great many people begin to ask themselves this question, a revolutionary guerrilla situation is incipient.

Introduction

A revolutionary war is never confined within the bounds of military action. Because its purpose is to destroy an existing society and its institutions and to replace them with a completely new state structure, any revolutionary war is a unity of which the constituent parts, in varying importance, are military, political, economic, social, and psychological. For this reason, it is endowed with a dynamic quality and a dimension in depth that orthodox wars, whatever their scale, lack. This is particularly true of revolutionary guerrilla war, which is not susceptible to the type of superficial military treatment frequently advocated by antediluvian doctrinaires.

It is often said that guerrilla warfare is primitive. This generalization is dangerously misleading and true only in the technological sense. If one considers the picture as a whole, a paradox is immediately apparent, and the primitive form is understood to be in fact more sophisticated than nuclear war or atomic war or war as it was waged by conventional armies, navies, and air forces. Guerrilla war is not dependent for success on the efficient operation of complex mechanical devices, highly organized logistical systems, or the accuracy of electronic computers. It can be conducted in any terrain, in any climate, in any weather; in swamps, in mountains, in farmed fields. Its basic element is man, and man is more complex than any of his machines. He is endowed with intelligence, emotions, and will. Guerrilla warfare is therefore suffused with, and reflects, man's admirable qualities as well as his less pleasant ones. While it is not always humane, it is human, which is more than can be said for the strategy of extinction.

In the United States, we go to considerable trouble to keep soldiers out of politics, and even more to keep politics out of soldiers. Guerrillas do exactly the opposite. They go to great lengths to make sure that their men are politically educated and thoroughly aware of the issues at stake. A trained and disciplined guerrilla is much more than a patriotic peasant, workman, or student armed with an antiquated fowling-piece and a home-made bomb. His indoctrination begins even before he is taught to shoot accurately, and it is unceasing. The end product is an intensely loyal and politically alert fighting man.

Guerrilla leaders spend a great deal more time in organization, instruction, agitation, and propaganda work than they do fighting, for their most important job is to win over the people. "We must patiently explain," says Mao Tse-tung. "Explain," "persuade," "discuss," "convince"—these words recur with monotonous regularity in many of the early Chinese essays on guerrilla war. Mao has aptly compared guerrillas to fish, and the people to the water in which they swim. If the political temperature is right, the fish, however few in number, will thrive and proliferate. It is therefore the principal concern of all guerrilla leaders to get the water to the right temperature and to keep it there.

More than ten years ago, I concluded an analysis of guerrilla warfare with the suggestion that the problem urgently demanded further "serious study of all historical experience." Although a wealth of material existed then, and much more has since been developed, no such study has yet been undertaken in this country, so far as I am aware. In Indochina and Cuba, Ho Chi Minh and Ernesto (Che) Guevara were more assiduous. One rather interesting result of their successful activities has been the common identification of guerrilla warfare with Communism. But guerrilla warfare was not invented by the Communists; for centuries, there have been guerrilla fighters.

One of the most accomplished of them all was our own Revolutionary hero Francis Marion, "the Swamp Fox." Those present at his birth would probably not have foretold a martial future for him; the baby was "not larger than a New England lobster and might easily enough have been put into a quart pot." Marion grew up in South Carolina and had little formal schooling. He worked as a farmer. In 1759, at the age of twenty-seven, he joined a regiment raised to fight the Cherokees, who were then ravaging the borders of the Carolinas. He served for two years and in the course of these hostilities stored away in his mind much that was later to be put to good use against the British.

When the Revolution broke out, Marion immediately accepted a commission in the Second South Carolina Regiment. By 1780, he had seen enough of the war to realize that the Continentals were overlooking a very profitable field—that of partisan warfare. Accordingly, he sought and obtained permission to organize a company that at first consisted of twenty ill-equipped men and boys (Castro's "base" was twelve men). The appearance of this group, with a heterogeneous assortment of arms and ragged and poorly fitting clothes, provoked considerable jesting among the regulars of General Gates, but Marion's men were not long in proving that the appearance of a combat soldier is not necessarily a reliable criterion of his fighting abilities.

Marion's guerrilla activities in South Carolina soon told heavily on the British, especially Cornwallis, whose plans were continually disrupted by them. Marion's tactics were those of all successful guerrillas. Operating with the greatest speed from inaccessible bases, which he changed frequently, he struck his blows in rapid succession at isolated garrisons, convoys, and trains. His information was always timely and accurate, for the people supported him.

The British, unable to cope with Marion, branded him a criminal, and complained bitterly that he fought neither "like a gentleman" nor like "a Christian," a charge orthodox soldiers are wont to apply in all lands and in all wars to such ubiquitous, intangible, and deadly antagonists as Francis Marion.*

However, the first example of guerrilla operations on a grand scale was in Spain between 1808 and 1813. The Spaniards who fled from Napoleon's invading army to the

---

* Bryant, in the "Song of Marion's Men," wrote some lines that showed that he had a better understanding of guerrilla tactics and psychology than many who have followed more martial pursuits:

> Woe to the English soldiery,
> That little dreads us near!
> On them shall come at midnight
> A strange and sudden fear;
> When, waking to their tents on fire,
> They grasp their arms in vain,
> And they who stand to face us
> Are beat to earth again;
> And they who fly in terror deem
> A mighty host behind,
> And hear the tramp of thousands
> Upon the hollow wind.

mountains were patriots loyal to the ruler whose crown had been taken from him by the Emperor of the French. They were not revolutionists. Most did not desire a change in the form of their government. Their single objective was to help Wellington force the French armies to leave Spain.

A few years later, thousands of Russian Cossacks and peasants harried Napoleon's Grande Armée as Kutuzov pushed it, stumbling, starving and freezing, down the ice-covered road to Smolensk. This dying army felt again and again the cudgel of the people's war, which, as Tolstoi later wrote, "was raised in all its menacing and majestic power; and troubling itself about no question of anyone's tastes or rules, about no fine distinctions, with stupid simplicity, with perfect consistency, it rose and fell and belabored the French until the whole invading army had been driven out."

A little more than a century and a quarter later, Hitler's armies fell back along the Smolensk road. They too would feel the fury of an aroused people. But in neither case were those who wielded the cudgel revolutionists. They were patriotic Russians.

Only when Lenin came on the scene did guerrilla warfare receive the potent political injection that was to alter its character radically. But it remained for Mao Tse-tung to produce the first systematic study of the subject, almost twenty-five years ago. His study, now endowed with the authority that deservedly accrues to the works of the man who led the most radical revolution in history, will continue to have a decisive effect in societies ready for change.

II

PROFILE OF A REVOLUTIONIST

Political power comes out of the barrel of a gun.
—MAO TSE-TUNG, 1938

MAO TSE-TUNG, the man who was to don the mantle of Lenin, was born in Hunan Province, in central China, in 1893. His father, an industrious farmer, had managed to acquire several acres, and with this land, the status of a "middle" peasant. He was a strict disciplinarian, and Mao's youth was not a happy one. The boy was in constant conflict with his father but found an ally in his mother, whose "indirect tactics" (as he once described her methods of coping with her husband) appealed to him. But the father gave his rebellious son educational opportunities that only a tiny minority of Chinese were then able to enjoy. Mao's primary and secondary schooling was thorough. His literary taste was catholic; while a pupil at the provincial normal school he read omnivorously. His indiscriminate diet included Chinese philosophy, poetry, history, and romances as well as translations of many Western historians, novelists, and biographers. However, history and political sciences particularly appealed to him; in them, he sought, but without success, the key to the future of China.

His studies had led him to reject both democratic liberalism and parliamentary socialism as unsuited to his country. Time, he realized, was running out for China. History would not accord her the privilege of gradual political, social, and economic change, of a relatively painless and orderly evolution. To survive in the power jungle, China had to change, to change radically, to change fast. But how?

Shortly after graduating from normal school, in 1917, Mao accepted a position as assistant in the Peking University library. Here he associated himself with the Marxist study groups set up by Li Ta-chao and Ch'en Tu-hsiu; here he discovered Lenin, read his essays, pored over Trotsky's explosive speeches, and began to study Marx and Engels. By 1920, Mao was a convinced Communist and a man who had discovered his mission: to create a new China according to the doctrine of Marx and Lenin. When the CCP was organized in Shanghai, in 1921, Mao joined.

The China Mao decided to change was not a nation in the accepted sense of the word. Culturally, China was, of course, homogeneous; politically and economically, China was chaos. The peasants, 400 million of them, lived from day to day at subsistence level. Tens of millions of peasant families owned no land at all. Other millions cultivated tiny holdings from which they scraped out just enough food to sustain life.

The peasant was fair game for everyone. He was pillaged by tax collectors, robbed by landlords and usurers, at the mercy of rapacious soldiery and bandits, afflicted by blights, droughts, floods, and epidemics. His single stark problem was simply to survive. The tough ones did. The others slowly starved, died of disease, and in the fierce winters of North China and Manchuria, froze to death.

It is difficult for an American today to conceive tens of thousands of small communities in which no public services existed, in which there were no doctors, no schools, no running water, no electricity, no paved streets, and no sewage disposal. The inhabitants of these communities were with few exceptions illiterate; they lived in constant fear of army press gangs and of provincial officials who called them out summer and winter alike to work on military roads and dikes. The Chinese peasant, in his own expressive idiom, "ate bitterness" from the time he could walk until he was laid to rest in the burial plot beneath the cypress trees. This was feudal China. Dormant within this society were the ingredients that were soon to blow it to pieces.

An external factor had for almost a century contributed to the chaos of China: the unrelenting pressure and greed of foreign powers. French, British, Germans, and Russians vied with one another in exacting from a succession of corrupt and feeble governments commercial, juridical, and financial concessions that had, in fact, turned China into an international colony. (The American record in these respects was a reasonably good one.) Mao once described the China he knew in his youth as "semicolonial and feudal." He was right.

Shortly after Chiang Kai-shek took command of the National Revolutionary Army, in 1926, Mao went to Hunan to stir up the peasants. The campaign he waged for land reform in his native province can be described as almost a one-man show. The fundamental requisite in China was then, as it had long been, to solve the land question. Reduced to elementary terms, the problem was how to get rid of the gentry landowners who fastened themselves to the peasants like leeches and whose exactions kept the people constantly impoverished. In the circumstances, there was only one way to accomplish this necessary reform: expropriation and redistribution of the land. Naturally, the Nationalists, eager to retain the support of the gentry (historically the stabilizing element in Chinese society), considered such a radical solution social dynamite. But in Mao's view, there could be no meaningful revolution unless and until the power of this class had been completely eliminated.

While Mao was making himself extremely unpopular with the landed gentry in Hunan, the revolutionary armies of the Kuomintang were marching north from Canton to Wuhan, on the Yangtze, where a Nationalist Government was established in December, 1926. These armies incorporated a number of Communist elements. But by the time the vanguard divisions of Chiang's army reached the outskirts of Shanghai, in March, 1927, the honeymoon was almost over. In April, Chiang's secret police captured and executed the radical labor leaders in Shanghai and began to purge the army of its Communist elements. In the meantime the left-wing government in Wuhan had broken up. The Communists walked out; the Soviet advisers packed their bags and started for home.

During this period, the Communists were having their own troubles, and these were serious. The movement was literally on the verge of extinction. Those who managed to escape Chiang's secret police had fled to the south and assembled at Ching Kang Shan, a rugged area in the Fukien-Kiangsi borderlands. One of the first to reach this haven was the agrarian agitator from Hunan. As various groups drifted in to the mountain stronghold, Mao and Chu Teh (who had arrived in April, 1928) began to mold an army. Several local bandit chieftains were induced to join the Communists, whose operations gradually became more extensive. Principally these activities were of a propaganda nature. District soviets were established; landlords were dispossessed; wealthy merchants were "asked" to make patriotic contributions. Gradually, the territory under Red control expanded, and from a temporarily secure base area, operations commenced against provincial troops who were supposed to suppress the Reds.

In the early summer of 1930, an ominous directive was received at Ching Kang Shan from the Central Committee of the Party, then dominated by Li Li-san. This directive required the Communist armies to take the offensive against cities held by the Nationalists. The campaigns that followed were not entirely successful and culminated in a serious Communist defeat at Changsha in September. On the thirteenth of that month, the single most vital decision in the history of the Chinese Communist Party was taken; the ultimate responsibility for it rested equally on the shoulders of Mao and Chu Teh. These two agreed that the only hope for the movement was to abandon immediately the line laid down by Moscow in favor of one of Mao's own devising. Basically the conflict that split the Chinese Communist Party wide open and alienated the traditionalists in Moscow revolved about this question: Was the Chinese revolution to be based on the industrial proletariat—as Marxist dogma prescribed—or was it to be based on the peasant? Mao, who knew and trusted the peasants, and had correctly gauged their revolutionary potential, was convinced that the Chinese urban proletariat were too few in number and too apathetic to make a revolution. This decision, which drastically reoriented the policy of the Chinese Communist Party, was thereafter to be carried out with vigorous consistency. History has proved that Mao was right, Moscow wrong. And it is for this reason that the doctrine of Kremlin infallibility is so frequently challenged by Peking.

In October, 1930, the Generalissimo, in the misguided belief that he could crush the Communists with no difficulty, announced with great fanfare a "Bandit Suppression Campaign." This was launched in December. How weak the Nationalists really were was now to become apparent. The campaign was a complete flop. Government troops ran away or surrendered to the Communists by platoons, by companies, by battalions. Three more Suppression Campaigns, all failures, followed this fiasco. Finally, in 1933, the Generalissimo reluctantly decided to adopt the plans of his German advisers and to commit well-equipped, well-trained, and loyal "Central" divisions to a coordinated and methodical compression of the Communist-controlled area. As the Nationalists inched southward, supported by artillery and aviation, they evacuated peasants from every village and town and constructed hundreds of mutually supporting wired-in blockhouses. The Communists, isolated from the support of the peasants they had laboriously converted, found themselves for the first time almost completely deprived of food and information. Chiang's troops were slowly strangling the Communists. For the first time, Communist morale sagged. It was in this context that the bold decision to shift the base to Shensi Province was taken, and the now celebrated march of almost 6,000 miles was begun.

This was indeed one of the fateful migrations of history: its purpose, to preserve the military power of the Communist Party. How many pitched battles and skirmishes the Reds fought during this epic trek cannot now be established. It is known, however, that for days on end their columns were under air attack. They crossed innumerable mountains and rivers and endured both tropical and subarctic climates. As they marched toward the borders of Tibet and swung north, they sprinkled the route with cadres and caches of arms and ammunition.

The Reds faced many critical situations, but they were tough and determined. Every natural obstacle, and there were many, was overcome. Chiang's provincial troops, ineffective as usual, were unable to bar the way, and the exhausted remnants of the Reds eventually found shelter in the loess caves of Pao An.

Later, after the base was shifted to Yenan, Mao had time to reflect on his experiences and to derive from them the theory and doctrine of revolutionary guerrilla war which he embodied in *Yu Chi Chan*.

III

STRATEGY, TACTICS, AND LOGISTICS IN REVOLUTIONARY WAR

> The first law of war is to preserve ourselves and destroy the enemy.
> —MAO TSE-TUNG, 1937

MAO HAS NEVER CLAIMED that guerrilla action alone is decisive in a struggle for political control of the state, but only that it is a possible, natural, and necessary development in an agrarian-based revolutionary war.

Mao conceived this type of war as passing through a series of merging phases, the first of which is devoted to organization, consolidation, and preservation of regional base areas situated in isolated and difficult terrain. Here volunteers are trained and indoctrinated, and from here, agitators and propagandists set forth, individually or in groups of two or three, to "persuade" and "convince" the inhabitants of the surrounding countryside and to enlist their support. In effect, there is thus woven about each base a protective belt of sympathizers willing to supply food, recruits, and information. The pattern of the process is conspiratorial, clandestine, methodical, and progressive. Military operations will be sporadic.

In the next phase, direct action assumes an ever-increasing importance. Acts of sabotage and terrorism multiply; collaborationists and "reactionary elements" are liquidated. Attacks are made on vulnerable military and police outposts; weak columns are ambushed. The primary purpose of these operations is to procure arms, ammunition, and other essential material, particularly medical supplies and radios. As the growing guerrilla force becomes better equipped and its capabilities improve, political agents proceed with indoctrination of the inhabitants of peripheral districts soon to be absorbed into the expanding "liberated" area.

One of the primary objectives during the first phases is to persuade as many people as possible to commit themselves to the movement, so that it gradually acquires the quality of "mass." Local "home guards" or militia are formed. The militia is not primarily designed to be a mobile fighting force; it is a "back-up" for the better-trained and better-equipped guerrillas. The home guards form an indoctrinated and partially trained reserve. They function as vigilantes. They collect information, force merchants to make "voluntary" contributions, kidnap particularly obnoxious local landlords, and liquidate informers and collaborators. Their function is to protect the revolution.

Following Phase I (organization, consolidation, and preservation) and Phase II (progressive expansion) comes Phase III: decision, or destruction of the enemy. It is during this period that a significant percentage of the active guerrilla force completes its transformation into an orthodox establishment capable of engaging the enemy in conventional battle. This phase may be protracted by "negotiations." Such negotiations are not originated by revolutionists for the purpose of arriving at amicable arrangements with the opposition. Revolutions rarely compromise; compromises are made only to further the strategic design. Negotiation, then, is undertaken for the dual purpose of gaining time to buttress a position (military, political, social, economic) and to wear down, frustrate, and harass the opponent. Few, if any, essential concessions are to be expected from the revolutionary side, whose aim is only to create conditions that will preserve the unity of the strategic line and guarantee the development of a "victorious situation."

Intelligence is the decisive factor in planning guerrilla operations. Where is the enemy? In what strength? What does he propose to do? What is the state of his equipment, his supply, his morale? Are his leaders intelligent, bold, and imaginative or stupid and impetuous? Are his troops tough, efficient, and well disciplined, or poorly trained and soft? Guerrillas expect the members of their intelligence service to provide the answers to these and dozens more detailed questions.

Guerrilla intelligence nets are tightly organized and pervasive. In a guerrilla area, every person without exception must be considered an agent—old men and women, boys driving ox carts, girls tending goats, farm laborers, storekeepers, schoolteachers, priests, boatmen, scavengers.

The local cadres "put the heat" on everyone, without regard to age or sex, to produce all conceivable information. And produce it they do.

As a corollary, guerrillas deny all information of themselves to their enemy, who is enveloped in an impenetrable fog. Total inability to get information was a constant complaint of the Nationalists during the first four Suppression Campaigns, as it was later of the Japanese in China and of the French in both Indochina and Algeria. This is a characteristic feature of all guerrilla wars. The enemy stands as on a lighted stage; from the darkness around him, thousands of unseen eyes intently study his every move, his every gesture. When he strikes out, he hits the air; his antagonists are insubstantial, as intangible as fleeting shadows in the moonlight.

Because of superior information, guerrillas always engage under conditions of their own choosing; because of superior knowledge of terrain, they are able to use it to their advantage and the enemy's discomfiture. Guerrillas fight only when the chances of victory are weighted heavily in their favor; if the tide of battle unexpectedly flows against them, they withdraw. They rely on imaginative leadership, distraction, surprise, and mobility to create a victorious situation before battle is joined. The enemy is deceived and again deceived. Attacks are sudden, sharp, vicious, and of short duration. Many are harassing in nature; others designed to dislocate the enemy's plans and to agitate and confuse his commanders. The mind of the enemy and the will of his leaders is a target of far more importance than the bodies of his troops. Mao once remarked, not entirely facetiously, that guerrillas must be expert at running away since they do it so often. They avoid static dispositions; their effort is always to keep the situation as fluid as possible, to strike where and when the enemy least expects them. Only in this way can they retain the initiative and so be assured of freedom of action. Usually designed to lure the enemy into a baited trap, to confuse his leadership, or to distract his attention from an area in which a more decisive blow is imminent, "running away" is thus, paradoxically, offensive.

Guerrilla operations conducted over a wide region are necessarily decentralized. Each regional commander must be familiar with local conditions and take advantage of local opportunities. The same applies to commands in subordinate districts. This decentralization is to some extent forced upon guerrillas because they ordinarily lack a well-developed system of technical communications. But at the same time, decentralization for normal operations has many advantages, particularly if local leaders are ingenious and bold.

The enemy's rear is the guerrillas' front; they themselves have no rear. Their logistical problems are solved in a direct and elementary fashion: The enemy is the principal source of weapons, equipment, and ammunition.

Mao once said:

We have a claim on the output of the arsenals of London as well as of Hanyang, and what is more, it is to be delivered to us by the enemy's own transport corps. This is the sober truth, not a joke.

If it is a joke, it is a macabre one as far as American tax-payers are concerned. Defectors to the Communists from Chiang Kai-shek's American-equipped divisions were numbered in the tens of thousands. When they surrendered, they turned in mountains of American-made individual arms, jeeps, tanks, guns, bazookas, mortars, radios, and automatic weapons.

It is interesting to examine Mao's strategical and tactical theories in the light of his principle of "unity of opposites." This seems to be an adaptation to military action of the ancient Chinese philosophical concept of *Yin-Yang.* Briefly, the *Yin* and the *Yang* are elemental and pervasive. Of opposite polarities, they represent female and male, dark and light, cold and heat, recession and aggression. Their reciprocal interaction is endless. In terms of the dialectic, they may be likened to the thesis and antithesis from which the synthesis is derived.

An important postulate of the *Yin-Yang* theory is that concealed within strength there is weakness, and within weakness, strength. It is a weakness of guerrillas that they operate in small groups that can be wiped out in a matter of minutes. But because they do operate in small groups, they can move rapidly and secretly into the vulnerable rear of the enemy.

In conventional tactics, dispersion of forces invites destruction; in guerrilla war, this very tactic is desirable both to confuse the enemy and to preserve the illusion that the guerrillas are ubiquitous.

It is often a disadvantage not to have heavy infantry

weapons available, but the very fact of having to transport them has until recently tied conventional columns to roads and well-used tracks. The guerrilla travels light and travels fast. He turns the hazards of terrain to his advantage and makes an ally of tropical rains, heavy snow, intense heat, and freezing cold. Long night marches are difficult and dangerous, but the darkness shields his approach to an unsuspecting enemy.

In every apparent disadvantage, some advantage is to be found. The converse is equally true: In each apparent advantage lie the seeds of disadvantage. The *Yin* is not wholly *Yin,* nor the *Yang* wholly *Yang.* It is only the wise general, said the ancient Chinese military philosopher Sun Tzu, who is able to recognize this fact and to turn it to good account.

Guerrilla tactical doctrine may be summarized in four Chinese characters pronounced "*Sheng Tung, Chi Hsi,*" which mean "Uproar [in the] East; Strike [in the] West." Here we find expressed the all-important principles of distraction on the one hand and concentration on the other; to fix the enemy's attention and to strike where and when he least anticipates the blow.

Guerrillas are masters of the arts of simulation and dissimulation; they create pretenses and simultaneously disguise or conceal their true semblance. Their tactical concepts, dynamic and flexible, are not cut to any particular pattern. But Mao's first law of war, to preserve oneself and destroy the enemy, is always governing.

IV

SOME CONCLUSIONS

Historical experience is written in blood and iron.
—Mao Tse-tung, 1937

THE FUNDAMENTAL DIFFERENCE between patriotic
partisan resistance and revolutionary guerrilla
movements is that the first usually lacks the ideological
content that always distinguishes the second.

A resistance is characterized by the quality of spon-
taneity; it begins and then is organized. A revolutionary
guerrilla movement is organized and then begins.

A resistance is rarely liquidated and terminates when
the invader is ejected; a revolutionary movement terminates
only when it has succeeded in displacing the incumbent
government or is liquidated.

Historical experience suggests that there is very little
hope of destroying a revolutionary guerrilla movement
*after it has survived the first phase and has acquired the
sympathetic support of a significant segment of the popu-
.lation.* The size of this "significant segment" will vary; a
decisive figure might range from 15 to 25 per cent.

In addition to an appealing program and popular support, such factors as terrain; communications; the quality

of the opposing leadership; the presence or absence of material help, technical aid, advisers, or "volunteers" from outside sources; the availability of a sanctuary; the relative military efficiency and the political flexibility of the incumbent government are naturally relevant to the ability of a movement to survive and expand.

In specific aspects, revolutionary guerrilla situations will of course differ, but if the Castro movement, for example, had been objectively analyzed in the light of the factors suggested during the latter period of its first phase, a rough "expectation of survival and growth" might have looked something like Figure I.

Had an impartial analyst applied such criteria to Vietnam six to eight months before the final debacle, he might have produced a chart somewhat like Figure II.

Here Determinants A, B, H, and I definitely favored the guerrillas, who also (unlike Castro) had an available sanctuary. Two others, C and F, might have been considered in balance. Although the Vietminh had demonstrated superior tactical ability in guerrilla situations, an experienced observer might have been justified in considering "military efficiency" equal; the French were learning.

While other determinants may no doubt be adduced, those used are, I believe, valid so far as they go, and the box scores indicative. These show that Castro's chances of success might have been estimated as approximately three to two, Ho Chi Minh's as approximately four to three.

These analyses may be criticized as having been formulated after the event; it is, however, my belief that the outcome in Cuba and Indochina could have been pre-

Figure I. The Revolutionary Guerrilla Situation in Cuba

Determinants*	Castro	Incumbent (Batista)	Remarks
A. Appeal of program	Progressive, plus (8)	Static, minus (3)	Batista government oppressive and reactionary
B. Popular support	Growing, active (7)	Diminishing, passive (3)	
C. Quality of leadership	Excellent, dedicated (8)	Mediocre to poor (4)	
D. Quality of troops	Good, improving to excellent (8)	Good, decreasing to fair (5)	
E. Military efficiency	Growing (6)	Mediocre to poor (4)	In guerrilla situations
F. Internal unity	Positive, strong (8)	Weak (3)	
G. Equipment	Poor, improving to good as taken (4)	Largely U.S., excellent (8)	Radios, transport, medical supplies, etc., available from incumbent
H. Base area terrain	Operationally favorable (10)	Unfavorable (3)	
I. Base area communications	Operationally favorable (10)	Unfavorable (3)	
J. Sanctuary	None (0)	Remainder of island (10)	Available for rest, retraining, equipment
AGGREGATE	69	46	

* Determinants are arbitrarily weighted on a scale of 0–10.

FIGURE II. THE REVOLUTIONARY GUERRILLA SITUATION IN VIETNAM

Determinants*	Ho Chi Minh	Incumbent (French)	Remarks
A. Appeal of program	Dynamic (7)	No program (0)	
B. Popular support	Growing (7)	Diminishing, slight (3)	
C. Quality of leadership	Good (7)	Good (7)	
D. Quality of troops	Good, improving (6)	Very good (7)	
E. Military efficiency	Very good (8)	Good (6)	In guerrilla situations
F. Internal unity	Excellent (8)	Excellent (8)	
G. Equipment	Fair but improving (7)	Generally well equipped (9)	Received from China and taken from French
H. Operational terrain	Favorable (10)	Unfavorable (5)	
I. Operational area communications	Favorable (10)	Unfavorable (5)	
J. Sanctuary	Available in China (8)	Remainder of Indochina (10)	
AGGREGATE	78	60	

* Determinants are arbitrarily weighted on a scale of 0–10.

dicted some time before the respective movements had emerged from the stage of organization and consolidation—Phase I.

At the present time, much attention is being devoted to the development of "gadgetry." A good example of this restricted approach to the problem was reported in *Newsweek:*[*]

> **PENTAGON**—A new and fiendishly ingenious antiguerrilla weapon is being tested by the Navy. It's a delayed-action liquid explosive, squirted from a flame-thrower-like gun, that seeps into foxholes and bunkers. Seconds later, fed by oxygen from the air, it blows up with terrific force.

Apparently we are to assume that guerrillas will conveniently ensconce themselves in readily identifiable "foxholes and bunkers" awaiting the arrival of half a dozen admirals armed with "flame-thrower-like guns" to march up, squirt, and retire to the nearest officers' club. To anyone even remotely acquainted with the philosophy and doctrine of revolutionary guerrilla war, this sort of thing is not hilariously funny. There are no mechanical panaceas.

I do not mean to suggest that proper weapons and equipment will not play an important part in antiguerrilla operations, for of course they will. Constant efforts should be made to improve communication, food, medical, and surgical "packs." Weapons and ammunition must be drastically reduced in weight; there seems to be no technical reason why a sturdy, light, accurate automatic rifle weigh-

[*] July 3, 1961, "The Periscope."

ing a maximum of four to five pounds cannot be developed. And the search for new and effective weapons must continue. But we must realize that "flame-thrower-like guns" and bullets are only a very small part of the answer to a challenging and complex problem.

The position of active third parties in a revolutionary guerrilla war and the timing, nature, and scope of the assistance given to one side or the other has become of great importance. Basically, this is a political matter; responsibility for a decision to intervene would naturally devolve upon the head of state. Any assistance given should, however, stop short of participation in combat. The role of a third party should be restricted to advice, materials, and technical training.

The timing of aid is often critical. If extended to the incumbent government, aid must be given while it is still possible to isolate and eradicate the movement; if to the revolutionary side, aid must be made available during the same critical period, that is, when the movement is vulnerable and its existence quite literally a matter of life and death.

From a purely military point of view, antiguerrilla operations may be summed up in three words: location, isolation, and eradication. In the brief definitions of each term, it will be well to bear in mind that these activities are not rigidly compartmented.

Location of base area or areas requires careful terrain studies, photographic and physical reconnaissance, and possibly infiltration of the movement. *Isolation* involves separation of guerrillas from their sources of information and food. It may require movement and resettlement of entire communities. *Eradication* presupposes reliable information and demands extreme operational flexibility and a high degree of mobility. Parachutists and helicopter-borne commando-type troops are essential.

The tactics of guerrillas must be used against the guerrillas themselves. They must be constantly harried and constantly attacked. Every effort must be made to induce defections and take prisoners. The best source of information of the enemy is men who know the enemy situation.

Imaginative, intelligent, and bold leadership is absolutely essential. *Commanders and leaders at every echelon* must be selected with these specific qualities in mind. Officers and NCO's who are more than competent under normal conditions will frequently be hopelessly ineffective when confronted with the dynamic and totally different situations characteristic of guerrilla warfare.

Finally, there is the question of whether it is possible to create effective counterguerrilla forces. Can two shoals of fish, each intent on destruction of the other, flourish in the same medium? Mao is definite on this point, he is convinced they cannot, that "counterrevolutionary guerrilla war" is impossible. If the guerrilla experiences of the White Russians (which he cites) or of Mikhailovitch are valid criteria, he is correct. But, on the other hand, the history of the movement in Greece during the German occupation indicates that under certain circumstances, his thesis will not stand too close an examination. This suggests the need for a careful analysis of relevant political factors in each individual situation.

Mao Tse-tung contends that the phenomena we have considered are subject to their own peculiar laws, and are predictable. If he is correct (and I believe he is), it is possible to prevent such phenomena from appearing, or, if they do, to control and eradicate them. And if historical experience teaches us anything about revolutionary guerrilla war, it is that military measures alone will not suffice.

YU CHI CHAN
(Guerrilla Warfare)

TRANSLATOR'S NOTE

In July, 1941, the undeclared war between China and Japan will enter its fifth year. One of the most significant features of the struggle has been the organization of the Chinese people for unlimited guerrilla warfare. The development of this warfare has followed the pattern laid out by Mao Tse-tung and his collaborators in the pamphlet *Yu Chi Chan (Guerrilla Warfare)*, which was published in 1937 and has been widely distributed in "Free China" at 10 cents a copy.

Mao Tse-tung, a member of the Chinese Communist Party and formerly political commissar of the Fourth Red Army, is no novice in the art of war. Actual battle experience with both regular and guerrilla troops has qualified him as an expert.

The influence of the ancient military philosopher Sun Tzu on Mao's military thought will be apparent to those who have read *The Book of War*. Sun Tzu wrote that speed, surprise, and deception were the primary essentials of the attack and his succinct advice, *"Sheng Tung, Chi Hsi"* ("Uproar [in the] East, Strike [in the] West"), is no less valid today than it was when he wrote it 2,400 years ago. The tactics of Sun Tzu are in large measure the tactics of China's guerrillas today.

Mao says that unlimited guerrilla warfare, with vast time and space factors, established a new military process. This seems a true statement since there are no other historical examples of guerrilla hostilities as thoroughly organized from the military, political, and economic point of view as those in China. We in the Marine Corps have as yet encountered nothing but relatively primitive and strictly limited guerrilla war. Thus, what Mao has written of this new type of guerrilla war may be of interest to us.

I have tried to present the author's ideas accurately, but as the Chinese language is not a particularly suitable medium for the expression of technical thought, the translation of some of the modern idioms not yet to be found in available dictionaries is probably arguable. I cannot vouch for the accuracy of retranslated quotations. I have taken the liberty to delete from the translation matter that was purely repetitious.

SAMUEL B. GRIFFITH
Captain, USMC

Quantico, Virginia
1940

A FURTHER NOTE

T HE PRECEDING NOTE was written twenty-one years ago, but I see no need to amplify it.

Yu Chi Chan (1937) is frequently confused with one of Mao's later (1938) essays entitled *K'ang Jih Yu Chi Chan Cheng Ti Chan Lueh Wen T'i (Strategic Problems in the Anti-Japanese Guerrilla War)*, which was issued in an English version in 1952 by the People's Publishing House, Peking. There are some similarities in these two works.

I had hoped to locate a copy of *Yu Chi Chan* in the Chinese to check my translation but have been unable to do so. Some improvement is always possible in any rendering from the Chinese. I have not been able to identify with standard English titles all the works cited by Mao.

Mao wrote *Yu Chi Chan* during China's struggle against Japan; consequently there are, naturally, numerous references to the strategy to be used against the Japanese. These in no way invalidate Mao's fundamental thesis. For instance, when Mao writes, "The moment that this war of resistance dissociates itself from the masses of the people is the precise moment that it dissociates itself from hope of ultimate victory over the Japanese," he might have added, "and from hope of ultimate victory over the forces of Chiang Kai-shek." However, he did not do so, because at that time both sides were attempting to preserve the illusion of a "united front." "Our basic policy," he said, "is the creation of a national united anti-Japanese front." This was, of course, *not* the basic policy of the Chinese Communist Party then, or at any other time. Its basic policy was to seize state power; the type of revolutionary guerrilla war described by Mao was the basic weapon in the protracted and ultimately successful process of doing so.

SAMUEL B. GRIFFITH
Brigadier General, USMC (Ret.)

Mount Vernon, Maine
July, 1961

1

WHAT IS GUERRILLA WARFARE?

IN A WAR OF REVOLUTIONARY CHARACTER, guerrilla operations are a necessary part. This is particularly true in a war waged for the emancipation of a people who inhabit a vast nation. China is such a nation, a nation whose techniques are undeveloped and whose communications are poor. She finds herself confronted with a strong and victorious Japanese imperialism. Under these circumstances, the development of the type of guerrilla warfare characterized by the quality of mass is both necessary and natural. This warfare must be developed to an unprecedented degree and it must coordinate with the operations of our regular armies. If we fail to do this, we will find it difficult to defeat the enemy.

These guerrilla operations must not be considered as an independent form of warfare. They are but one step in the total war, one aspect of the revolutionary struggle. They are the inevitable result of the clash between oppressor and oppressed when the latter reach the limits of their endurance. In our case, these hostilities began at a time when the people were unable to endure any more from the Japanese imperialists. Lenin, in *People and Revolution*, said: "A people's insurrection and a people's revolution are not only natural but inevitable." We consider guerrilla operations as but one aspect of our total or mass war because they, lacking the quality of independence, are of themselves incapable of providing a solution to the struggle.

Guerrilla warfare has qualities and objectives peculiar to itself. It is a weapon that a nation inferior in arms and military equipment may employ against a more powerful aggressor nation. When the invader pierces deep into the heart of the weaker country and occupies her territory in a cruel and oppressive manner, there is no doubt that conditions of terrain, climate, and society in general offer obstacles to his progress and may be used to advantage by those who oppose him. In guerrilla warfare, we turn these advantages to the purpose of resisting and defeating the enemy.

During the progress of hostilities, guerrillas gradually develop into orthodox forces that operate in conjunction with other units of the regular army. Thus the regularly organized troops, those guerrillas who have attained that status, and those who have not reached that level of development combine to form the military power of a national revolutionary war. There can be no doubt that the ultimate result of this will be victory.

Both in its development and in its method of application, guerrilla warfare has certain distinctive characteristics. We first discuss the relationship of guerrilla warfare to national policy. Because ours is the resistance of a semicolonial country against an imperialism, our hostilities must have a clearly defined political goal and firmly established political responsibilities. Our basic policy is the creation of a national united anti-Japanese front. This policy we pursue in order to gain our political goal, which is the complete emancipation of the Chinese people. There are certain fundamental steps necessary in the realization of this policy, to wit:

1. Arousing and organizing the people.
2. Achieving internal unification politically.
3. Establishing bases.
4. Equipping forces.
5. Recovering national strength.
6. Destroying enemy's national strength.
7. Regaining lost territories.

There is no reason to consider guerrilla warfare separately from national policy. On the contrary, it must be organized and conducted in complete accord with national anti-Japanese policy. It is only those who misinterpret guerrilla action who say, as does Jen Ch'i Shan, "The question of guerrilla hostilities is purely a military matter and not a political one." Those who maintain this simple point of view have lost sight of the political goal and the political effects of guerrilla action. Such a simple point of view will cause the people to lose confidence and will result in our defeat.

What is the relationship of guerrilla warfare to the people? Without a political goal, guerrilla warfare must fail, as it must if its political objectives do not coincide with the aspirations of the people and their sympathy, cooperation, and assistance cannot be gained. The essence of guerrilla warfare is thus revolutionary in character. On the other

43

hand, in a war of counterrevolutionary nature, there is no place for guerrilla hostilities. Because guerrilla warfare basically derives from the masses and is supported by them, it can neither exist nor flourish if it separates itself from their sympathies and cooperation. There are those who do not comprehend guerrilla action, and who therefore do not understand the distinguishing qualities of a people's guerrilla war, who say: "Only regular troops can carry on guerrilla operations." There are others who, because they do not believe in the ultimate success of guerrilla action, mistakenly say: "Guerrilla warfare is an insignificant and highly specialized type of operation in which there is no place for the masses of the people" (Jen Ch'i Shan). Then there are those who ridicule the masses and undermine resistance by wildly asserting that the people have no understanding of the war of resistance (Yeh Ch'ing, for one). The moment that this war of resistance dissociates itself from the masses of the people is the precise moment that it dissociates itself from hope of ultimate victory over the Japanese.

What is the organization for guerrilla warfare? Though all guerrilla bands that spring from the masses of the people suffer from lack of organization at the time of their formation, they all have in common a basic quality that makes organization possible. All guerrilla units must have political and military leadership. This is true regardless of the source or size of such units. Such units may originate locally, in the masses of the people; they may be formed from an admixture of regular troops with groups of the people, or they may consist of regular army units intact.

And mere quantity does not affect this matter. Such units may consist of a squad of a few men, a battalion of several hundred men, or a regiment of several thousand men.

All these must have leaders who are unyielding in their policies—resolute, loyal, sincere, and robust. These men must be well educated in revolutionary technique, self-confident, able to establish severe discipline, and able to cope with counterpropaganda. In short, these leaders must be models for the people. As the war progresses, such leaders will gradually overcome the lack of discipline, which at first prevails; they will establish discipline in their forces, strengthening them and increasing their combat efficiency. Thus eventual victory will be attained.

Unorganized guerrilla warfare cannot contribute to victory and those who attack the movement as a combination of banditry and anarchism do not understand the nature of guerrilla action. They say: "This movement is a haven for disappointed militarists, vagabonds and bandits" (Jen Ch'i Shan), hoping thus to bring the movement into disrepute. We do not deny that there are corrupt guerrillas, nor that there are people who under the guise of guerrillas indulge in unlawful activities. Neither do we deny that the movement has at the present time symptoms of a lack of organization, symptoms that might indeed be serious were we to judge guerrilla warfare solely by the corrupt and temporary phenomena we have mentioned. We should study the corrupt phenomena and attempt to eradicate them in order to encourage guerrilla warfare, and to increase its military efficiency. "This is hard work, there is no help for it, and the problem cannot be solved immediately. The whole people must try to reform themselves during the course of the war. We must educate them and reform them in the light of past experience. Evil does not exist in guerrilla warfare but only in the unorganized and undisciplined activities that are anarchism," said Lenin, in *On Guerrilla Warfare.**

What is basic guerrilla strategy? Guerrilla strategy must be based primarily on alertness, mobility, and attack. It must be adjusted to the enemy situation, the terrain, the existing lines of communication, the relative strengths, the weather, and the situation of the people.

In guerrilla warfare, select the tactic of seeming to come from the east and attacking from the west; avoid the solid, attack the hollow; attack; withdraw; deliver a lightning blow, seek a lightning decision. When guerrillas engage a stronger enemy, they withdraw when he advances; harass him when he stops; strike him when he is weary; pursue him when he withdraws. In guerrilla strategy, the enemy's rear, flanks, and other vulnerable spots are his vital points, and there he must be harassed, attacked, dispersed, exhausted and annihilated. Only in this way can guerrillas carry out their mission of independent guerrilla action and coordination with the effort of the regular armies. But, in spite of the most complete preparation, there can be no victory if mistakes are made in the matter of command. Guerrilla warfare based on the principles we have mentioned and carried on over a vast extent of territory in which

* Presumably, Mao refers here to the essay that has been translated into English under the title "Partisan Warfare." See *Orbis*, II (Summer, 1958), No. 2, 194–208.–S.B.G.

communications are inconvenient will contribute tremendously towards ultimate defeat of the Japanese and consequent emancipation of the Chinese people.

A careful distinction must be made between two types of guerrilla warfare. The fact that revolutionary guerrilla warfare is based on the masses of the people does not in itself mean that the organization of guerrilla units is impossible in a war of counterrevolutionary character. As examples of the former type we may cite Red guerrilla hostilities during the Russian Revolution; those of the Reds in China; of the Abyssinians against the Italians for the past three years; those of the last seven years in Manchuria, and the vast anti-Japanese guerrilla war that is carried on in China today. All these struggles have been carried on in the interests of the whole people or the greater part of them; all had a broad basis in the national manpower, and all have been in accord with the laws of historical development. They have existed and will continue to exist, flourish, and develop as long as they are not contrary to national policy.

The second type of guerrilla warfare directly contradicts the law of historical development. Of this type, we may cite the examples furnished by the White Russian guerrilla units organized by Denikin and Kolchak; those organized by the Japanese; those organized by the Italians in Abyssinia; those supported by the puppet governments in Manchuria and Mongolia, and those that will be organized here by Chinese traitors. All such have oppressed the masses and have been contrary to the true interests of the people. They must be firmly opposed. They are easy to destroy because they lack a broad foundation in the people.

If we fail to differentiate between the two types of guerrilla hostilities mentioned, it is likely that we will exaggerate their effect when applied by an invader. We might arrive at the conclusion that "the invader can organize guerrilla units from among the people." Such a conclusion might well diminish our confidence in guerrilla warfare. As far as this matter is concerned, we have but to remember the historical experience of revolutionary struggles.

Further, we must distinguish general revolutionary wars from those of a purely "class" type. In the former case, the whole people of a nation, without regard to class or party, carry on a guerrilla struggle that is an instrument of the national policy. Its basis is, therefore, much broader than is the basis of a struggle of class type. Of a general guerrilla war, it has been said: "When a nation is invaded, the people become sympathetic to one another and all aid in organizing guerrilla units. In civil war, no matter to what extent guerrillas are developed, they do not produce the same results as when they are formed to resist an invasion by foreigners" (*Civil War in Russia*).* The one strong feature of guerrilla warfare in a civil struggle is its quality of internal purity. One class may be easily united and perhaps fight with great effect, whereas in a national revolutionary war, guerrilla units are faced with the problem of internal unification of different class groups. This necessitates the use of propaganda. Both types of guerrilla

---

* Presumably, Mao refers here to *Lessons of Civil War*, by S. I. Gusev; first published in 1918 by the Staff Armed Forces, Ukraine; revised in 1921 and published by GIZ, Moscow; reprinted in 1958 by the Military Publishing House, Moscow.—S.B.G.

war are, however, similar in that they both employ the same military methods.

National guerrilla warfare, though historically of the same consistency, has employed varying implements as times, peoples, and conditions differ. The guerrilla aspects of the Opium War, those of the fighting in Manchuria since the Mukden incident, and those employed in China today are all slightly different. The guerrilla warfare conducted by the Moroccans against the French and the Spanish was not exactly similar to that which we conduct today in China. These differences express the characteristics of different peoples in different periods. Although there is a general similarity in the quality of all these struggles, there are dissimilarities in form. This fact we must recognize. Clausewitz wrote, in *On War:* "Wars in every period have independent forms and independent conditions, and, therefore, every period must have its independent theory of war." Lenin, in *On Guerrilla Warfare,* said: "As regards the form of fighting, it is unconditionally requisite that history be investigated in order to discover the conditions of environment, the state of economic progress, and the political ideas that obtained, the national characteristics, customs, and degree of civilization." Again: "It is necessary to be completely unsympathetic to abstract formulas and rules and to study with sympathy the conditions of the actual fighting, for these will change in accordance with the political and economic situations and the realization of the people's aspirations. These progressive changes in conditions create new methods."

If, in today's struggle, we fail to apply the historical

truths of revolutionary guerrilla war, we will fall into the error of believing with T'ou Hsi Sheng that under the impact of Japan's mechanized army, "the guerrilla unit has lost its historical function." Jen Ch'i Shan writes: "In olden days, guerrilla warfare was part of regular strategy but there is almost no chance that it can be applied today." These opinions are harmful. If we do not make an estimate of the characteristics peculiar to our anti-Japanese guerrilla war, but insist on applying to it mechanical formulas derived from past history, we are making the mistake of placing our hostilities in the same category as all other national guerrilla struggles. If we hold this view, we will simply be beating our heads against a stone wall and we will be unable to profit from guerrilla hostilities.

To summarize: What is the guerrilla war of resistance against Japan? It is one aspect of the entire war, which, although alone incapable of producing the decision, attacks the enemy in every quarter, diminishes the extent of area under his control, increases our national strength, and assists our regular armies. It is one of the strategic instruments used to inflict defeat on our enemy. It is the one pure expression of anti-Japanese policy, that is to say, it is military strength organized by the active people and inseparable from them. It is a powerful special weapon with which we resist the Japanese and without which we cannot defeat them.

2

THE RELATION OF GUERRILLA
HOSTILITIES TO REGULAR
OPERATIONS

T HE GENERAL FEATURES of orthodox hostilities,
that is, the war of position and the war of movement, differ fundamentally from guerrilla warfare. There
are other readily apparent differences such as those in
organization, armament, equipment, supply, tactics, command; in conception of the terms "front" and "rear"; in the
matter of military responsibilities.

When considered from the point of view of total numbers, guerrilla units are many; as individual combat units,
they may vary in size from the smallest, of several score
or several hundred men, to the battalion or the regiment, of
several thousand. This is not the case in regularly organized units. A primary feature of guerrilla operations is their
dependence upon the people themselves to organize battalions and other units. As a result of this, organization
depends largely upon local circumstances. In the case of
guerrilla groups, the standard of equipment is of a low
order, and they must depend for their sustenance primarily
upon what the locality affords.

The strategy of guerrilla warfare is manifestly unlike that employed in orthodox operations, as the basic tactic of the former is constant activity and movement. There is in guerrilla warfare no such thing as a decisive battle; there is nothing comparable to the fixed, passive defense that characterizes orthodox war. In guerrilla warfare, the transformation of a moving situation into a positional defensive situation never arises. The general features of reconnaissance, partial deployment, general deployment, and development of the attack that are usual in mobile warfare are not common in guerrilla war.

There are differences also in the matter of leadership and command. In guerrilla warfare, small units acting independently play the principal role, and there must be no excessive interference with their activities. In orthodox warfare, particularly in a moving situation, a certain degree of initiative is accorded subordinates, but in principle, command is centralized. This is done because all units and all supporting arms in all districts must coordinate to the highest degree. In the case of guerrilla warfare, this is not only undesirable but impossible. Only adjacent guerrilla units can coordinate their activities to any degree. Strategically, their activities can be roughly correlated with those of the regular forces, and tactically, they must cooperate with adjacent units of the regular army. But there are no strictures on the extent of guerrilla activity nor is it primarily characterized by the quality of cooperation of many units.

When we discuss the terms "front" and "rear," it must be remembered, that while guerrillas do have bases, their primary field of activity is in the enemy's rear areas. They themselves have no rear. Because an orthodox army has rear installations (except in some special cases as during the 10,000-mile* march of the Red Army or as in the case of certain units operating in Shansi Province), it cannot operate as guerrillas can.

As to the matter of military responsibilities, those of the guerrillas are to exterminate small forces of the enemy; to harass and weaken large forces; to attack enemy lines of communication; to establish bases capable of supporting independent operations in the enemy's rear; to force the enemy to disperse his strength; and to coordinate all these activities with those of the regular armies on distant battle fronts.

From the foregoing summary of differences that exist between guerrilla warfare and orthodox warfare, it can be seen that it is improper to compare the two. Further distinction must be made in order to clarify this matter. While the Eighth Route Army is a regular army, its North China campaign is essentially guerrilla in nature, for it operates in the enemy's rear. On occasion, however, Eighth Route Army commanders have concentrated powerful forces to strike an enemy in motion, and the characteristics of orthodox mobile warfare were evident in the battle at P'ing Hsing Kuan and in other engagements.

On the other hand, after the fall of Feng Ling Tu, the operations of Central Shansi, and Suiyuan, troops were more guerrilla than orthodox in nature. In this connection,

* It has been estimated that the Reds actually marched about 6,000 miles. See Introduction, Chapter II.—S.B.G.

the precise character of Generalissimo Chiang's instructions to the effect that independent brigades would carry out guerrilla operations should be recalled. In spite of such temporary activities, these orthodox units retained their identity and after the fall of Feng Ling Tu, they not only were able to fight along orthodox lines but often found it necessary to do so. This is an example of the fact that orthodox armies may, due to changes in the situation, temporarily function as guerrillas.

Likewise, guerrilla units formed from the people may gradually develop into regular units and, when operating as such, employ the tactics of orthodox mobile war. While these units function as guerrillas, they may be compared to innumerable gnats, which, by biting a giant both in front and in rear, ultimately exhaust him. They make themselves as unendurable as a group of cruel and hateful devils, and as they grow and attain gigantic proportions, they will find that their victim is not only exhausted but practically perishing. It is for this very reason that our guerrilla activities are a source of constant mental worry to Imperial Japan.

While it is improper to confuse orthodox with guerrilla operations, it is equally improper to consider that there is a chasm between the two. While differences do exist, similarities appear under certain conditions, and this fact must be appreciated if we wish to establish clearly the relationship between the two. If we consider both types of warfare as a single subject, or if we confuse guerrilla warfare with the mobile operations of orthodox war, we fall into this error: We exaggerate the function of guerrillas and minimize that of the regular armies. If we agree with Chang Tso Hua, who says, "Guerrilla warfare is the primary war strategy of a people seeking to emancipate itself," or with Kao Kang, who believes that "Guerrilla strategy is the only strategy possible for an oppressed people," we are exaggerating the importance of guerrilla hostilities. What these zealous friends I have just quoted do not realize is this: If we do not fit guerrilla operations into their proper niche, we cannot promote them realistically. Then, not only would those who oppose us take advantage of our varying opinions to turn them to their own uses to undermine us, but guerrillas would be led to assume responsibilities they could not successfully discharge and that should properly be carried out by orthodox forces. In the meantime, the important guerrilla function of coordinating activities with the regular forces would be neglected.

Furthermore, if the theory that guerrilla warfare is our only strategy were actually applied, the regular forces would be weakened, we would be divided in purpose, and guerrilla hostilities would decline. If we say, "Let us transform the regular forces into guerrillas," and do not place our first reliance on a victory to be gained by the regular armies over the enemy, we may certainly expect to see as a result the failure of the anti-Japanese war of resistance. The concept that guerrilla warfare is an end in itself and that guerrilla activities can be divorced from those of the regular forces is incorrect. If we assume that guerrilla warfare does not progress from beginning to end beyond its elementary forms, we have failed to recognize the fact that guerrilla hostilities can, under specific conditions, develop and assume orthodox characteristics. An opinion that admits the existence of guerrilla war, but isolates it, is one that does not properly estimate the potentialities of such war.

Equally dangerous is the concept that condemns guerrilla war on the ground that war has no other aspects than the purely orthodox. This opinion is often expressed by those who have seen the corrupt phenomena of some guerrilla regimes, observed their lack of discipline, and have seen them used as a screen behind which certain persons have indulged in bribery and other corrupt practices. These people will not admit the fundamental necessity for guerrilla bands that spring from the armed people. They say, "Only the regular forces are capable of conducting guerrilla operations." This theory is a mistaken one and would lead to the abolition of the people's guerrilla war.

A proper conception of the relationship that exists between guerrilla effort and that of the regular forces is essential. We believe it can be stated this way: "Guerrilla operations during the anti-Japanese war may for a certain time and temporarily become its paramount feature, particularly insofar as the enemy's rear is concerned. However, if we view the war as a whole, there can be no doubt that our regular forces are of primary importance, because it is they who are alone capable of producing the decision. Guerrilla warfare assists them in producing this favorable decision. Orthodox forces may under certain conditions operate as guerrillas, and the latter may, under certain conditions, develop to the status of the former. However, both guerrilla forces and regular forces have their own respective development and their proper combinations."

To clarify the relationship between the mobile aspect of orthodox war and guerrilla war, we may say that general agreement exists that the principal element of our strategy must be mobility. With the war of movement, we may at times combine the war of position. Both of these are assisted by general guerrilla hostilities. It is true that on the battlefield mobile war often becomes positional; it is true that this situation may be reversed; it is equally true that each form may combine with the other. The possibility of such combination will become more evident after the prevailing standards of equipment have been raised. For example, in a general strategical counterattack to recapture key cities and lines of communication, it would be normal to use both mobile and positional methods. However, the point must again be made that our fundamental strategical form must be the war of movement. If we deny this, we cannot arrive at the victorious solution of the war. In sum, while we must promote guerrilla warfare as a necessary strategical auxiliary to orthodox operations, we must neither assign it the primary position in our war strategy nor substitute it for mobile and positional warfare as conducted by orthodox forces.

3

GUERRILLA WARFARE IN HISTORY

G UERRILLA WARFARE is neither a product of
China nor peculiar to the present day. From
the earliest historical days, it has been a feature of wars
fought by every class of men against invaders and oppres-
sors. Under suitable conditions, it has great possibilities.
The many guerrilla wars in history have their points of
difference, their peculiar characteristics, their varying proc-
esses and conclusions, and we must respect and profit by
the experience of those whose blood was shed in them.
What a pity it is that the priceless experience gained during the several hundred wars waged by the peasants of
China cannot be marshaled today to guide us. Our only
experience in guerrilla hostilities has been that gained
from the several conflicts that have been carried on against
us by foreign imperialisms. But that experience should
help the fighting Chinese recognize the necessity for guerrilla warfare and should confirm them in confidence of
ultimate victory.

In September, 1812, the Frenchman Napoleon, in the
course of swallowing all of Europe, invaded Russia at the
head of a great army totaling several hundred thousand
infantry, cavalry, and artillery. At that time, Russia was

weak and her ill-prepared army was not concentrated. The most important phase of her strategy was the use made of Cossack cavalry and detachments of peasants to carry on guerrilla operations. After giving up Moscow, the Russians formed nine guerrilla divisions of about five hundred men each. These, and vast groups of organized peasants, carried on partisan warfare and continually harassed the French Army. When the French Army was withdrawing, cold and starving, Russian guerrillas blocked the way and, in combination with regular troops, carried out counterattacks on the French rear, pursuing and defeating them. The army of the heroic Napoleon was almost entirely annihilated, and the guerrillas captured many officers, men, cannon, and rifles. Though the victory was the result of various factors, and depended largely on the activities of the regular army, the function of the partisan groups was extremely important. "The corrupt and poorly organized country that was Russia defeated and destroyed an army led by the most famous soldier of Europe and won the war in spite of the fact that her ability to organize guerrilla regimes was not fully developed. At times, guerrilla groups were hindered in their operations and the supply of equipment and arms was insufficient. If we use the Russian saying, it was a case of a battle between 'the fist and the ax' " (Ivanov).

From 1918 to 1920, the Russian Soviets, because of the opposition and intervention of foreign imperialisms and the internal disturbances of White Russian groups, were forced to organize themselves in occupied territories and fight a real war. In Siberia and Alashan, in the rear of the army of the traitor Denikin and in the rear of the Poles, there were many Red Russian guerrillas. These not only disrupted and destroyed the communications in the enemy's rear but also frequently prevented his advance. On one occasion, the guerrillas completely destroyed a retreating White Army that had previously been defeated by regular Red forces. Kolchak, Denikin, the Japanese, and the Poles, owing to the necessity of staving off the attacks of guerrillas, were forced to withdraw regular troops from the front. "Thus not only was the enemy's manpower impoverished but he found himself unable to cope with the ever-moving guerrilla" (*The Nature of Guerrilla Action*).

The development of guerrillas at that time had only reached the stage where there were detached groups of several thousands in strength, old, middle aged, and young. The old men organized themselves into propaganda groups known as "silver-haired units"; there was a suitable guerrilla activity for the middle aged; the young men formed combat units, and there were even groups for the children. Among the leaders were determined Communists who carried on general political work among the people. These, although they opposed the doctrine of extreme guerrilla warfare, were quick to oppose those who condemned it. Experience tells us that "Orthodox armies are the fundamental and principal power; guerrilla units are secondary to them and assist in the accomplishment of the mission assigned the regular forces" (*Lessons of the Civil War in Russia*).* Many of the guerrilla regimes in Russia gradually developed until in battle they were able to dis-

* See p. 48 n.—S.B.G.

charge functions of organized regulars. The army of the famous General Galen was entirely derived from guerrillas.

During seven months in 1935 and 1936, the Abyssinians lost their war against Italy. The cause of defeat—aside from the most important political reasons that there were dissentient political groups, no strong government party, and unstable policy—was the failure to adopt a positive policy of mobile warfare. There was never a combination of the war of movement with large-scale guerrilla operations. Ultimately, the Abyssinians adopted a purely passive defense, with the result that they were unable to defeat the Italians. In addition to this, the fact that Abyssinia is a relatively small and sparsely populated country was contributory. Even in spite of the fact that the Abyssinian Army and its equipment were not modern, she was able to withstand a mechanized Italian force of 400,000 for seven months. During that period, there were several occasions when a war of movement was combined with large-scale guerrilla operations to strike the Italians heavy blows. Moreover, several cities were retaken and casualties totaling 140,000 were inflicted. Had this policy been steadfastly continued, it would have been difficult to have named the ultimate winner. At the present time, guerrilla activities continue in Abyssinia, and if the internal political questions can be solved, an extension of such activities is probable.

In 1841 and 1842, when brave people from San Yuan Li fought the English; again from 1850 to 1864, during the Taiping War, and for a third time in 1899, in the Boxer Uprising, guerrilla tactics were employed to a remarkable degree. Particularly was this so during the Taiping War, when guerrilla operations were most extensive and the Ch'ing troops were often completely exhausted and forced to flee for their lives.

In these wars, there were no guiding principles of guerrilla action. Perhaps these guerrilla hostilities were not carried out in conjunction with regular operations, or perhaps there was a lack of coordination. But the fact that victory was not gained was not because of any lack in guerrilla activity but rather because of the interference of politics in military affairs. Experience shows that if precedence is not given to the question of conquering the enemy in both political and military affairs, and if regular hostilities are not conducted with tenacity, guerrilla operations alone cannot produce final victory.

From 1927 to 1936, the Chinese Red Army fought almost continually and employed guerrilla tactics constantly. At the very beginning, a positive policy was adopted. Many bases were established, and from guerrilla bands, the Reds were able to develop into regular armies. As these armies fought, new guerrilla regimes were developed over a wide area. These regimes coordinated their efforts with those of the regular forces. This policy accounted for the many victories gained by guerrilla troops relatively few in number, who were armed with weapons inferior to those of their opponents. The leaders of that period properly combined guerrilla operations with a war of movement both strategically and tactically. They depended primarily upon alertness. They stressed the correct basis for both political affairs and military operations. They developed their guerrilla bands into trained units. They then determined upon a ten-year period of resistance during which time they overcame innumerable difficulties and have only lately reached their goal of direct participation in the anti-Japanese war. There is no doubt that the internal unification of China is now a permanent and definite fact and that the experience gained during our internal struggles has proved to be both necessary and advantageous to us in the struggle against Japanese imperialism. There are many valuable lessons we can learn from the experience of those years. Principal among them is the fact that guerrilla success largely depends upon powerful political leaders who work unceasingly to bring about internal unification. Such leaders must work with the people; they must have a correct conception of the policy to be adopted as regards both the people and the enemy.

After September 18, 1931, strong anti-Japanese guerrilla campaigns were opened in each of the three northeast provinces. Guerrilla activity persists there in spite of the cruelties and deceits practiced by the Japanese at the expense of the people, and in spite of the fact that her armies have occupied the land and oppressed the people for the last seven years. The struggle can be divided into two periods. During the first, which extended from September 18, 1931, to January, 1933, anti-Japanese guerrilla activity exploded constantly in all three provinces. Ma Chan Shan and Ssu Ping Wei established an anti-Japanese regime in Heilungkiang. In Chi Lin, the National Salvation Army and the Self-Defense Army were led by Wang Te Lin and Li Tu respectively. In Feng T'ien, Chu Lu and others commanded guerrilla units. The influence of these forces was great. They harassed the Japanese unceasingly, but because there was an indefinite political goal, improper leadership, failure to coordinate military command and operations and to work with the people, and, finally, failure to delegate proper political functions to the army, the whole organization was feeble, and its strength was not unified. As a direct result of these conditions, the campaigns failed and the troops were finally defeated by our enemy.

During the second period, which has extended from January, 1933, to the present time, the situation has greatly improved. This has come about because great numbers of people who have been oppressed by the enemy have decided to resist him, because of the participation of the Chinese Communists in the anti-Japanese war, and because of the fine work of the volunteer units. The guerrillas have finally educated the people to the meaning of guerrilla warfare, and in the northeast, it has again become an important and powerful influence. Already seven or eight guerrilla regiments and a number of independent platoons have been formed, and their activities make it necessary for the Japanese to send troops after them month after month. These units hamper the Japanese and undermine their control in the northeast, while, at the same time, they inspire a Nationalist revolution in Korea. Such activities are not merely of transient and local importance but directly contribute to our ultimate victory.

However, there are still some weak points. For instance: National defense policy has not been sufficiently developed;

participation of the people is not general; internal political organization is still in its primary stages, and the force used to attack the Japanese and the puppet governments is not yet sufficient. But if present policy is continued tenaciously, all these weaknesses will be overcome. Experience proves that guerrilla war will develop to even greater proportions and that, in spite of the cruelty of the Japanese and the many methods they have devised to cheat the people, they cannot extinguish guerrilla activities in the three northeastern provinces.

The guerrilla experiences of China and of other countries that have been outlined prove that in a war of revolutionary nature such hostilities are possible, natural and necessary. They prove that if the present anti-Japanese war for the emancipation of the masses of the Chinese people is to gain ultimate victory, such hostilities must expand tremendously.

Historical experience is written in iron and blood. We must point out that the guerrilla campaigns being waged in China today are a page in history that has no precedent. Their influence will not be confined solely to China in her present anti-Japanese war but will be world-wide.

4

CAN VICTORY BE ATTAINED BY GUERRILLA OPERATIONS?

GUERRILLA HOSTILITIES are but one phase of the war of resistance against Japan and the answer to the question of whether or not they can produce ultimate victory can be given only after investigation and comparison of all elements of our own strength with those of the enemy. The particulars of such a comparison are several. First, the strong Japanese bandit nation is an absolute monarchy. During the course of her invasion of China, she had made comparative progress in the techniques of industrial production and in the development of excellence and skill in her army, navy, and air force. But in spite of this industrial progress, she remains an absolute monarchy of inferior physical endowments. Her manpower, her raw materials, and her financial resources are all inadequate and insufficient to maintain her in protracted warfare or to meet the situation presented by a war prosecuted over a vast area. Added to this is the antiwar feeling now manifested by the Japanese people, a feeling that is shared by the junior officers and, more extensively, by the soldiers of the invading army. Furthermore, China is not Japan's only enemy. Japan is unable to employ her entire strength in the attack on China; she cannot, at most, spare more than a million men for this purpose, as she must hold any in excess of that number for use against other possible opponents. Because of these important primary considerations, the invading Japanese bandits can hope neither to be victorious in a protracted struggle nor to conquer a vast area. Their strategy must be one of lightning war and speedy decision. If we can hold out for three or more years, it will be most difficult for Japan to bear up under the strain.

In the war, the Japanese brigands must depend upon lines of communication linking the principal cities as routes for the transport of war materials. The most important considerations for her are that her rear be stable and peaceful and that her lines of communication be intact. It is not to her advantage to wage war over a vast area with disrupted lines of communication. She cannot disperse her strength and fight in a number of places, and her greatest fears are thus eruptions in her rear and disruption of her lines of communication. If she can maintain communications, she will be able at will to concentrate powerful forces speedily at strategic points to engage our organized units in decisive battle. Another important Japanese objective is to profit from the industries, finances, and manpower in captured areas and with them to augment her own insufficient strength. Certainly, it is not to her advantage to forgo these benefits, nor to be forced to dissipate her energies in a type of warfare in which the gains will not compensate for the losses. It is for these reasons that guerrilla warfare conducted in each bit of conquered territory over a wide area will be a heavy blow struck at the Japanese bandits. Experience in the five northern provinces as well as in Kiangsu, Chekiang, and Anhwei has absolutely established the truth of this assertion.

China is a country half colonial and half feudal; it is a country that is politically, militarily, and economically backward. This is an inescapable conclusion. It is a vast country with great resources and tremendous population, a country in which the terrain is complicated and the facilities for communication are poor. All these factors favor a protracted war; they all favor the application of mobile warfare and guerrilla operations. The establishment of innumerable anti-Japanese bases behind the enemy's lines will force him to fight unceasingly in many places at once, both to his front and his rear. He thus endlessly expends his resources.

We must unite the strength of the army with that of the people; we must strike the weak spots in the enemy's flanks, in his front, in his rear. We must make war everywhere and cause dispersal of his forces and dissipation of his strength. Thus the time will come when a gradual change will become evident in the relative position of ourselves and our enemy, and when that day comes, it will be the beginning of our ultimate victory over the Japanese.

Although China's population is great, it is unorganized. This is a weakness which must be taken into account.

The Japanese bandits have invaded our country not

merely to conquer territory but to carry out the violent, rapacious, and murderous policy of their government, which is the extinction of the Chinese race. For this compelling reason, we must unite the nation without regard to parties or classes and follow our policy of resistance to the end. China today is not the China of old. It is not like Abyssinia. China today is at the point of her greatest historical progress. The standards of literacy among the masses have been raised; the *rapprochement* of Communists and Nationalists has laid the foundation for an anti-Japanese war front that is constantly being strengthened and expanded; government, army, and people are all working with great energy; the raw-material resources and the economic strength of the nation are waiting to be used; the unorganized people is becoming an organized nation.

These energies must be directed toward the goal of protracted war so that should the Japanese occupy much of our territory or even most of it, we shall still gain final victory. Not only must those behind our lines organize for resistance but also those who live in Japanese-occupied territory in every part of the country. The traitors who accept the Japanese as fathers are few in number, and those who have taken oath that they would prefer death to abject slavery are many. If we resist with this spirit, what enemy can we not conquer and who can say that ultimate victory will not be ours?

The Japanese are waging a barbaric war along uncivilized lines. For that reason, Japanese of all classes oppose the policies of their government, as do vast international groups. On the other hand, because China's cause is right-eous, our countrymen of all classes and parties are united to oppose the invader; we have sympathy in many foreign countries, including even Japan itself. This is perhaps the most important reason why Japan will lose and China will win.

The progress of the war for the emancipation of the Chinese people will be in accord with these facts. The guerrilla war of resistance will be in accord with these facts, and that guerrilla operations correlated with those of our regular forces will produce victory is the conviction of the many patriots who devote their entire strength to guer-rilla hostilities.

5

ORGANIZATION FOR GUERRILLA WARFARE

FOUR POINTS MUST BE CONSIDERED under this subject. These are:

1. How are guerrilla bands formed?
2. How are guerrilla bands organized?
3. What are the methods of arming guerrilla bands?
4. What elements constitute a guerrilla band?

These are all questions pertaining to the organization of armed guerrilla units; they are questions which those who have had no experience in guerrilla hostilities do not understand and on which they can arrive at no sound decisions; indeed, they would not know in what manner to begin.

How Guerrilla Units Are Originally Formed

The unit may originate in any one of the following ways:

a) From the masses of the people.

b) From regular army units temporarily detailed for the purpose.

c) From regular army units permanently detailed.

d) From the combination of a regular army unit and a unit recruited from the people.

e) From the local militia.

f) From deserters from the ranks of the enemy.

g) From former bandits and bandit groups.

In the present hostilities, no doubt, all these sources will be employed.

In the first case above, the guerrilla unit is formed from the people. This is the fundamental type. Upon the arrival of the enemy army to oppress and slaughter the people, their leaders call upon them to resist. They assemble the most valorous elements, arm them with old rifles or bird guns, and thus a guerrilla unit begins. Orders have already been issued throughout the nation that call upon the people to form guerrilla units both for local defense and for other combat. If the local governments approve and aid such movements, they cannot fail to prosper. In some places, where the local government is not determined or where its officers have all fled, the leaders among the masses (relying on the sympathy of the people and their sincere desire to resist Japan and succor the country) call upon the people to resist, and they respond. Thus, many guerrilla units are organized. In circumstances of this kind, the duties of leadership usually fall upon the shoulders of young students, teachers, professors, other educators, local soldiery, professional men, artisans, and those without a fixed profession, who are willing to exert themselves to the last drop of their blood. Recently, in Shansi, Hopeh, Chahar, Suiyuan, Shantung, Chekiang, Anhwei, Kiangsu, and other provinces, extensive guerrilla hostilities have broken out. All these are organized and led by patriots. The amount of such activity is the best proof of the foregoing statement. The more such bands there are, the better will the situation be. Each district, each county, should be able to organize a great number of guerrilla squads, which, when assembled, form a guerrilla company.

There are those who say: "I am a farmer," or, "I am a student"; "I can discuss literature but not military arts." This is incorrect. There is no profound difference between the farmer and the soldier. You must have courage. You simply leave your farms and become soldiers. That you are farmers is of no difference, and if you have education, that is so much the better. When you take your arms in hand, you become soldiers; when you are organized, you become military units.

Guerrilla hostilities are the university of war, and after you have fought several times valiantly and aggressively, you may become a leader of troops, and there will be many well-known regular soldiers who will not be your peers. Without question, the fountainhead of guerrilla warfare is in the masses of the people, who organize guerrilla units directly from themselves.

The second type of guerrilla unit is that which is organized from small units of the regular forces temporarily detached for the purpose. For example, since hostilities commenced, many groups have been temporarily detached from armies, divisions, and brigades and have been assigned guerrilla duties. A regiment of the regular army may, if circumstances warrant, be dispersed into groups for the purpose of carrying on guerrilla operations. As an example of this, there is the Eighth Route Army, in North China. Excluding the periods when it carries on mobile operations as an army, it is divided into its elements and these carry on guerrilla hostilities. This type of guerrilla unit is essential for two reasons. First, in mobile-warfare situations, the coordination of guerrilla activities with regular operations is necessary. Second, until guerrilla hostilities can be developed on a grand scale, there is no one to carry out guerrilla missions but regulars. Historical experience shows us that regular army units are not able to undergo the hardships of guerrilla campaigning over long periods. The leaders of regular units engaged in guerrilla operations must be extremely adaptable. They must study the methods of guerrilla war. They must understand that initiative, discipline, and the employment of stratagems are all of the utmost importance. As the guerrilla status of regular units is but temporary, their leaders must lend all possible support to the organization of guerrilla units from among the people. These units must be so disciplined that they hold together after the departure of the regulars.

The third type of unit consists of a detachment of regulars who are permanently assigned guerrilla duties.. This type of small detachment does not have to be prepared to rejoin the regular forces. Its post is somewhere in the rear of the enemy, and there it becomes the backbone of guerrilla organization. As an example of this type of organization, we may take the Wu Tai Shan district in the heart of the Hopeh-Chahar-Shansi area. Along the borders of these provinces, units from the Eighth Route Army have established a framework for guerrilla operations. Around these small cores, many detachments have been organized and the area of guerrilla activity greatly expanded. In areas in which there is a possibility of cutting the enemy's lines of supply, this system should be used. Severing enemy supply routes destroys his life line; this is one feature that cannot be neglected. If, at the time the regular forces withdraw from a certain area, some units are left behind, these should conduct guerrilla operations in the enemy's rear. As an example of this, we have the guerrilla bands now continuing their independent operations in the Shanghai-Woosung area in spite of the withdrawal of regular forces.

The fourth type of organization is the result of a merger between small regular detachments and local guerrilla units. The regular forces may dispatch a squad, a platoon, or a company, which is placed at the disposal of the local guerrilla commander. If a small group experienced in military and political affairs is sent, it becomes the core of the local guerrilla unit. These several methods are all excellent, and if properly applied, the intensity of guerrilla warfare can be extended. In the Wu Tai Shan area, each of these methods has been used.

The fifth type mentioned above is formed from the local militia, from police and home guards. In every North China province, there are now many of these groups, and they should be formed in every locality. The government has issued a mandate to the effect that the people are not to depart from war areas. The officer in command of the county, the commander of the peace-preservation unit, the chief of police are all required to obey this mandate. They cannot retreat with their forces but must remain at their stations and resist.

The sixth type of unit is that organized from troops that come over from the enemy—the Chinese "traitor troops" employed by the Japanese. It is always possible to produce disaffection in their ranks, and we must increase our propaganda efforts and foment mutinies among such troops. Immediately after mutinying, they must be received into our ranks and organized. The concord of the leaders and the assent of the men must be gained, and the units rebuilt politically and reorganized militarily. Once this has been accomplished, they become successful guerrilla units. In regard to this type of unit, it may be said that political work among them is of the utmost importance.

The seventh type of guerrilla organization is that formed from bands of bandits and brigands. This, although difficult, must be carried out with utmost vigor lest the enemy use such bands to his own advantage. Many bandit groups pose as anti-Japanese guerrillas, and it is only necessary to correct their political beliefs to convert them.

In spite of inescapable differences in the fundamental types of guerrilla bands, it is possible to unite them to form a vast sea of guerrillas. The ancients said, "Tai Shan is a great mountain because it does not scorn the merest handful of dirt; the rivers and seas are deep because they absorb the waters of small streams." Attention paid to the enlistment and organization of guerrillas of every type and from every source will increase the potentialities of guerrilla action in the anti-Japanese war. This is something that patriots will not neglect.

Yu Chi Chan (Guerrilla Warfare)

THE METHOD OF ORGANIZING GUERRILLA REGIMES

Many of those who decide to participate in guerrilla activities do not know the methods of organization. For such people, as well as for students who have no knowledge of military affairs, the matter of organization is a problem that requires solution. Even among those who have military knowledge, there are some who know nothing of guerrilla regimes because they are lacking in that particular type of experience. The subject of the organization of such regimes is not confined to the organization of specific units but includes all guerrilla activities within the area where the regime functions.

As an example of such organization, we may take a geographical area in the enemy's rear. This area may comprise many counties. It must be subdivided and individual companies or battalions formed to accord with the subdivisions. To this "military area," a military commander and political commissioners are appointed. Under these, the necessary officers, both military and political, are appointed. In the military headquarters, there will be the staff, the aides, the supply officers, and the medical personnel. These are controlled by the chief of staff, who acts in accordance with orders from the commander. In the political headquarters, there are bureaus of propaganda organization, people's mass movements, and miscellaneous affairs. Control of these is vested in the political chairmen.

The military areas are subdivided into smaller districts in accordance with local geography, the enemy situation locally, and the state of guerrilla development. Each of these smaller divisions within the area is a district, each of which may consist of from two to six counties. To each district, a military commander and several political commissioners are appointed. Under their direction, military and political headquarters are organized. Tasks are assigned in accordance with the number of guerrilla troops available. Although the names of the officers in the "district" correspond to those in the larger "area," the number of functionaries assigned in the former case should be reduced to the least possible. In order to unify control, to handle guerrilla troops that come from different sources, and to harmonize military operations and local political affairs, a committee of from seven to nine members should be organized in each area and district. This committee, the members of which are selected by the troops and the local political officers, should function as a forum for the discussion of both military and political matters.

All the people in an area should arm themselves and be organized into two groups. One of these groups is a combat group, the other a self-defense unit with but limited military quality. Regular combatant guerrillas are organized into one of three general types of unit. The first of these is the small unit, the platoon or company. In each county, three to six units may be organized. The second type is the battalion of from two to four companies. One such unit should be organized in each county. While the unit fundamentally belongs to the county in which it was organized, it may operate in other counties. While in areas other than its own, it must operate in conjunction with local units in order to take advantage of their manpower, their knowledge of local terrain and local customs, and their information of the enemy.

The third type is the guerrilla regiment, which consists of from two to four of the above-mentioned battalion units. If sufficient manpower is available, a guerrilla brigade of from two to four regiments may be formed.

Each of the units has its own peculiarities of organization. A squad, the smallest unit, has a strength of from nine to eleven men, including the leader and the assistant leader. Its arms may be from two to five Western-style rifles, with the remaining men armed with rifles of local manufacture, bird guns, spears, or big swords. Two to four such squads form a platoon. This, too, has a leader and an assistant leader, and when acting independently, it is assigned a political officer to carry on political propaganda work. The platoon may have about ten rifles, with the remainder of its weapons being bird guns, lances, and big swords. Two to four of such units form a company, which, like the platoon, has a leader, an assistant leader, and a political officer. All these units are under the direct supervision of the military commanders of the areas in which they operate.

The battalion unit must be more thoroughly organized and better equipped than the smaller units. Its discipline and its personnel should be superior. If a battalion is formed from company units, it should not deprive subordinate units entirely of their manpower and their arms. If, in a small area, there is a peace-preservation corps, a branch of the militia, or police, regular guerrilla units should not be dispersed over it.

The guerrilla unit next in size to the battalion is the regiment. This must be under more severe discipline than the battalion. In an independent guerrilla regiment, there may be ten men per squad, three squads per platoon, three platoons per company, three companies per battalion, and three battalions to the regiment. Two of such regiments form a brigade. Each of these units has a commander, a vice-commander, and a political officer.

In North China, guerrilla cavalry units should be established. These may be regiments of from two to four companies, or battalions.

All these units from the lowest to the highest are combatant guerrilla units and receive their supplies from the central government. Details of their organization are shown in the tables.*

All the people of both sexes from the ages of sixteen to forty-five must be organized into anti-Japanese self-defense units, the basis of which is voluntary service. As a first step, they must procure arms, then they must be given both military and political training. Their responsibilities are: local sentry duties, securing information of the enemy, arresting traitors, and preventing the dissemination of enemy propaganda. When the enemy launches a guerrilla-suppression drive, these units, armed with what weapons there are, are assigned to certain areas to deceive, hinder, and harass him. Thus, the self-defense units assist the combatant guerrillas. They have other functions. They furnish stretcher-bearers to transport the wounded, carriers to take food to the troops, and comfort missions to provide

* See Appendix.—S.B.G.

the troops with tea and rice. If a locality can organize such a self-defense unit as we have described, the traitors cannot hide nor can bandits and robbers disturb the peace of the people. Thus the people will continue to assist the guerrillas and supply manpower to our regular armies. "The organization of self-defense units is a transitional step in the development of universal conscription. Such units are reservoirs of manpower for the orthodox forces."

There have been such organizations for some time in Shansi, Shensi, Honan, and Suiyuan. The youth organizations in different provinces were formed for the purpose of educating the young. They have been of some help. However, they were not voluntary, and the confidence of the people was thus not gained. These organizations were not widespread, and their effect was almost negligible. This system was, therefore, supplanted by the new-type organizations, which are organized on the principles of voluntary cooperation and nonseparation of the members from their native localities. When the members of these organizations are in their native towns, they support themselves. Only in case of military necessity are they ordered to remote places, and when this is done, the government must support them. Each member of these groups must have a weapon even if the weapon is only a knife, a pistol, a lance, or a spear.

In all places where the enemy operates, these self-defense units should organize within themselves a small guerrilla group of perhaps from three to ten men armed with pistols or revolvers. This group is not required to leave its native locality.

The organization of these self-defense units is mentioned in this book because such units are useful for the purposes of inculcating the people with military and political knowledge, keeping order in the rear, and replenishing the ranks of the regulars. These groups should be organized not only in the active war zones but in every province in China. "The people must be inspired to cooperate voluntarily. We must not force them, for if we do, it will be ineffectual." This is extremely important. The organization of a self-defense army similar to that we have mentioned is shown in Table 5.*

In order to control anti-Japanese military organization as a whole, it is necessary to establish a system of military areas and districts along the lines we have indicated. The organization of such areas and districts is shown in Table 6.

EQUIPMENT OF GUERRILLAS

In regard to the problem of guerrilla equipment, it must be understood that guerrillas are lightly armed attack groups, which require simple equipment. The standard of equipment is based upon the nature of duties assigned; the equipment of low-class guerrilla units is not as good as that of higher-class units. For example, those who are assigned the task of destroying railroads are better-equipped than those who do not have that task. The equipment of guerrillas cannot be based on what the guerrillas want, or even what they need, but must be based on what is available for their use. Equipment cannot be furnished

* Unfortunately, this table, as well as Table 6, was omitted from the edition of *Yu Chi Chan* available to me.—S.B.G.

immediately but must be acquired gradually. These are points to be kept in mind.

The question of equipment includes the collection, supply, distribution, and replacement of weapons, ammunition, blankets, communication materials, transport, and facilities for propaganda work. The supply of weapons and ammunition is most difficult, particularly at the time the unit is established, but this problem can always be solved eventually. Guerrilla bands that originate in the people are furnished with revolvers, pistols, bird guns, spears, big swords, and land mines and mortars of local manufacture. Other elementary weapons are added and as many new-type rifles as are available are distributed. After a period of resistance, it is possible to increase the supply of equipment by capturing it from the enemy. In this respect, the transport companies are the easiest to equip, for in any successful attack, we will capture the enemy's transport.

An armory should be established in each guerrilla district for the manufacture and repair of rifles and for the production of cartridges, hand grenades, and bayonets. Guerrillas must not depend too much on an armory. The enemy is the principal source of their supply.

For destruction of railway trackage, bridges, and stations in enemy-controlled territory, it is necessary to gather together demolition materials. Troops must be trained in the preparation and use of demolitions, and a demolition unit must be organized in each regiment.

As for minimum clothing requirements, these are that each man shall have at least two summer-weight uniforms, one suit of winter clothing, two hats, a pair of wrap put tees, and a blanket. Each man must have a haversack or a bag for food. In the north, each man must have an overcoat. In acquiring this clothing, we cannot depend on captures made from the enemy, for it is forbidden for captors to take clothing from their prisoners. In order to maintain high morale in guerrilla forces, all the clothing and equipment mentioned should be furnished by the representatives of the government stationed in each guerrilla district. These men may confiscate clothing from traitors or ask contributions from those best able to afford them. In subordinate groups, uniforms are unnecessary.

Telephone and radio equipment is not necessary in lower groups, but all units from regiment up are equipped with both. This material can be obtained by contributions from the regular forces and by capture from the enemy.

In the guerrilla army in general, and at bases in particular, there must be a high standard of medical equipment. Besides the services of the doctors, medicines must be procured. Although guerrillas can depend on the enemy for some portion of their medical supplies, they must, in general, depend upon contributions. If Western medicines are not available, local medicines must be made to suffice.

The problem of transport is more vital in North China than in the south, for in the south all that are necessary are mules and horses. Small guerrilla units need no animals, but regiments and brigades will find them necessary. Commanders and staffs of units from companies up should be furnished a riding animal each. At times, two officers will have to share a horse. Officers whose duties are of minor nature do not have to be mounted.

Propaganda materials are very important. Every large guerrilla unit should have a printing press and a mimeograph stone. They must also have paper on which to print propaganda leaflets and notices. They must be supplied with chalk and large brushes. In guerrilla areas, there should be a printing press or a lead-type press.

For the purpose of printing training instructions, this material is of the greatest importance.

In addition to the equipment listed above, it is necessary to have field glasses, compasses, and military maps. An accomplished guerrilla group will acquire these things.

Because of the proved importance of guerrilla hostilities in the anti-Japanese war, the headquarters of the Nationalist Government and the commanding officers of the various war zones should do their best to supply the guerrillas with what they actually need and are unable to get for themselves. However, it must be repeated that guerrilla equipment will in the main depend on the efforts of the guerrillas themselves. If they depend on higher officers too much, the psychological effect will be to weaken the guerrilla spirit of resistance.

ELEMENTS OF THE GUERRILLA ARMY

The term "element" as used in the title to this section refers to the personnel, both officers and men, of the guerrilla army. Since each guerrilla group fights in a protracted war, its officers must be brave and positive men whose entire loyalty is dedicated to the cause of emancipation of the people. An officer should have the following qualities: great powers of endurance so that in spite of any hardship he sets an example to his men and is a model for them; he must be able to mix easily with the people; his spirit and that of the men must be one in strengthening the policy of resistance to the Japanese. If he wishes to gain victories, he must study tactics. A guerrilla group with officers of this caliber would be unbeatable. I do not mean that every guerrilla group can have, at its inception, officers of such qualities. The officers must be men naturally endowed with good qualities which can be developed during the course of campaigning. The most important natural quality is that of complete loyalty to the idea of people's emancipation. If this is present, the others will develop; if it is not present, nothing can be done. When officers are first selected from a group, it is this quality that should receive particular attention. The officers in a group should be inhabitants of the locality in which the group is organized, as this will facilitate relations between them and the local civilians. In addition, officers so chosen would be familiar with conditions. If in any locality there are not enough men of sufficiently high qualifications to become officers, an effort must be made to train and educate the people so these qualities may be developed and the potential officer material increased. There can be no disagreements between officers native to one place and those from other localities.

A guerrilla group ought to operate on the principle that only volunteers are acceptable for service. It is a mistake to impress people into service. As long as a person is willing to fight, his social condition or position is no consideration, but only men who are courageous and determined can bear the hardships of guerrilla campaigning in a protracted war.

A soldier who habitually breaks regulations must be dismissed from the army. Vagabonds and vicious people must not be accepted for service. The opium habit must be forbidden, and a soldier who cannot break himself of the habit should be dismissed. Victory in guerrilla war is conditioned upon keeping the membership pure and clean.

It is a fact that during the war the enemy may take advantage of certain people who are lacking in conscience and patriotism and induce them to join the guerrillas for the purpose of betraying them. Officers must, therefore, continually educate the soldiers and inculcate patriotism in them. This will prevent the success of traitors. The traitors who are in the ranks must be discovered and expelled, and punishment and expulsion meted out to those who have been influenced by them. In all such cases, the officers should summon the soldiers and relate the facts to them, thus arousing their hatred and detestation for traitors. This procedure will serve as well as a warning to the other soldiers. If an officer is discovered to be a traitor, some prudence must be used in the punishment adjudged. However, the work of eliminating traitors in the army begins with their elimination from among the people.

Chinese soldiers who have served under puppet governments and bandits who have been converted should be welcomed as individuals or as groups. They should be well treated and repatriated. But care should be used during their reorientation to distinguish those whose idea is to fight the Japanese from those who may be present for other reasons.

6

THE POLITICAL PROBLEMS OF
GUERRILLA WARFARE

I N CHAPTER 1, I mentioned the fact that guerrilla
troops should have a precise conception of the
political goal of the struggle and the political organization
to be used in attaining that goal. This means that both
organization and discipline of guerrilla troops must be at a
high level so that they can carry out the political activities
that are the life of both the guerrilla armies and of revolutionary warfare.

First of all, political activities depend upon the indoctrination of both military and political leaders with the
idea of anti-Japanism. Through them, the idea is trans-
mitted to the troops. One must not feel that he is anti-
Japanese merely because he is a member of a guerrilla unit.
The anti-Japanese idea must be an ever-present conviction,
and if it is forgotten, we may succumb to the temptations
of the enemy or be overcome with discouragements. In a
war of long duration, those whose conviction that the people must be emancipated is not deep rooted are likely to
become shaken in their faith or actually revolt. Without
the general education that enables everyone to understand

our goal of driving out Japanese imperialism and establishing a free and happy China, the soldiers fight without conviction and lose their determination.

The political goal must be clearly and precisely indicated to inhabitants of guerrilla zones and their national consciousness awakened. Hence, a concrete explanation of the political systems used is important not only to guerrilla troops but to all those who are concerned with the realization of our political goal. The Kuomintang has issued a pamphlet entitled *System of National Organization for War*, which should be widely distributed throughout guerrilla zones. If we lack national organization, we will lack the essential unity that should exist between the soldiers and the people.

A study and comprehension of the political objectives of this war and of the anti-Japanese front is particularly important for officers of guerrilla troops. There are some militarists who say: "We are not interested in politics but only in the profession of arms." It is vital that these simple-minded militarists be made to realize the relationship that exists between politics and military affairs. Military action is a method used to attain a political goal. While military affairs and political affairs are not identical, it is impossible to isolate one from the other.

It is to be hoped that the world is in the last era of strife. The vast majority of human beings have already prepared or are preparing to fight a war that will bring justice to the oppressed peoples of the world. No matter how long this war may last, there is no doubt that it will be followed by an unprecedented epoch of peace. The war that we are fighting today for the emancipation of the Chinese is a part of the war for the freedom of all human beings, and the independent, happy, and liberal China that we are fighting to establish will be a part of that new world order. A conception like this is difficult for the simple-minded militarist to grasp and it must therefore be carefully explained to him.

There are three additional matters that must be considered under the broad question of political activities. These are political activities, first, as applied to the troops; second, as applied to the people; and, third, as applied to the enemy. The fundamental problems are: first, spiritual unification of officers and men within the army; second, spiritual unification of the army and the people; and, last, destruction of the unity of the enemy. The concrete methods for achieving these unities are discussed in detail in pamphlet Number 4 of this series, entitled *Political Activities in Anti-Japanese Guerrilla Warfare*.

A revolutionary army must have discipline that is established on a limited democratic basis. In all armies, obedience of the subordinates to their superiors must be exacted. This is true in the case of guerrilla discipline, but the basis for guerrilla discipline must be the individual conscience. With guerrillas, a discipline of compulsion is ineffective. In any revolutionary army, there is unity of purpose as far as both officers and men are concerned, and, therefore, within such an army, discipline is self-imposed. Although discipline in guerrilla ranks is not as severe as in the ranks of orthodox forces, the necessity for discipline exists. This must be self-imposed, because only when it is, is the soldier able to understand completely why he fights and why he must obey. This type of discipline becomes a tower of strength within the army, and it is the only type that can truly harmonize the relationship that exists between officers and soldiers.

In any system where discipline is externally imposed, the relationship that exists between officer and man is characterized by indifference of the one to the other. The idea that officers can physically beat or severely tongue-lash their men is a feudal one and is not in accord with the conception of a self-imposed discipline. Discipline of the feudal type will destroy internal unity and fighting strength. A discipline self-imposed is the primary characteristic of a democratic system in the army.

A secondary characteristic is found in the degree of liberties accorded officers and soldiers. In a revolutionary army, all individuals enjoy political liberty and the question, for example, of the emancipation of the people must not only be tolerated but discussed, and propaganda must be encouraged. Further, in such an army, the mode of living of the officers and the soldiers must not differ too much, and this is particularly true in the case of guerrilla troops. Officers should live under the same conditions as their men, for that is the only way in which they can gain from their men the admiration and confidence so vital in war. It is incorrect to hold to a theory of equality in all things, but there must be equality of existence in accepting the hardships and dangers of war. Thus we may attain to the unification of the officer and soldier groups, a unity both horizontal within the group itself, and vertical, that is, from lower to higher echelons. It is only when such unity is present that units can be said to be powerful combat factors.

There is also a unity of spirit that should exist between troops and local inhabitants. The Eighth Route Army put into practice a code known as "The Three Rules and the Eight Remarks," which we list here:

Rules:
1. All actions are subject to command.
2. Do not steal from the people.
3. Be neither selfish nor unjust.

Remarks:
1. Replace the door when you leave the house.*
2. Roll up the bedding on which you have slept.
3. Be courteous.
4. Be honest in your transactions.
5. Return what you borrow.
6. Replace what you break.
7. Do not bathe in the presence of women.
8. Do not without authority search the pocketbooks of those you arrest.

The Red Army adhered to this code for ten years and the Eighth Route Army and other units have since adopted it.

Many people think it impossible for guerrillas to exist for long in the enemy's rear. Such a belief reveals lack of com-

* In summer, doors were frequently lifted off and used as beds. —S.B.G.

prehension of the relationship that should exist between the people and the troops. The former may be likened to water and the latter to the fish who inhabit it. How may it be said that these two cannot exist together? It is only undisciplined troops who make the people their enemies and who, like the fish out of its native element, cannot live.

We further our mission of destroying the enemy by propagandizing his troops, by treating his captured soldiers with consideration, and by caring for those of his wounded who fall into our hands. If we fail in these respects, we strengthen the solidarity of our enemy.

7

THE STRATEGY OF GUERRILLA
RESISTANCE AGAINST JAPAN

I T HAS BEEN DEFINITELY DECIDED that in the strategy
of our war against Japan, guerrilla strategy must
be auxiliary to fundamental orthodox methods. If this were
a small country, guerrilla activities could be carried out
close to the scene of operations of the regular army and
directly complementary to them. In such a case, there would
be no question of guerrilla strategy as such. Nor would the
question arise if our country were as strong as Russia, for
example, and able speedily to eject an invader. The question exists because China, a weak country of vast size, has
today progressed to the point where it has become possible
to adopt the policy of a protracted war characterized by
guerrilla operations. Although these may at first glance
seem to be abnormal or heterodox, such is not actually the
case.

Because Japanese military power is inadequate, much of
the territory her armies have overrun is without sufficient
garrison troops. Under such circumstances the primary
functions of guerrillas are three: first, to conduct a war on
exterior lines, that is, in the rear of the enemy; second, to

establish bases; and, last, to extend the war areas. Thus, guerrilla participation in the war is not merely a matter of purely local guerrilla tactics but involves strategical considerations.

Such war, with its vast time and space factors, establishes a new military process, the focal point of which is China today. The Japanese are apparently attempting to recall a past that saw the Yüan extinguish the Sung and the Ch'ing conquer the Ming; that witnessed the extension of the British Empire to North America and India; that saw the Latins overrun Central and South America. As far as China today is concerned, such dreams of conquest are fantastic and without reality. Today's China is better equipped than was the China of yesterday, and a new type of guerrilla hostilities is a part of that equipment. If our enemy fails to take these facts into consideration and makes too optimistic an estimate of the situation, he courts disaster.

Though the strategy of guerrillas is inseparable from war strategy as a whole, the actual conduct of these hostilities differs from the conduct of orthodox operations. Each type of warfare has methods peculiar to itself, and methods suitable to regular warfare cannot be applied with success to the special situations that confront guerrillas.

Before we treat the practical aspects of guerrilla war, it might be well to recall the fundamental axiom of combat on which all military action is based. This can be stated: "Conservation of one's own strength; destruction of enemy strength." A military policy based on this axiom is consonant with a national policy directed towards the building of a free and prosperous Chinese state and the destruction of Japanese imperialism. It is in furtherance of this policy that government applies its military strength. Is the sacrifice demanded by war in conflict with the idea of self-preservation? Not at all. The sacrifices demanded are necessary both to destroy the enemy and to preserve ourselves; the sacrifice of a part of the people is necessary to preserve the whole. All the considerations of military action are derived from this axiom. Its application is as apparent in all tactical and strategical conceptions as it is in the simple case of the soldier who shoots at his enemy from a covered position.

All guerrilla units start from nothing and grow. What methods should we select to ensure the conservation and development of our own strength and the destruction of that of the enemy? The essential requirements are the six listed below:

1. Retention of the initiative; alertness; carefully planned tactical attacks in a war of strategical defense; tactical speed in a war strategically protracted; tactical operations on exterior lines in a war conducted strategically on interior lines.

2. Conduct of operations to complement those of the regular army.

3. The establishment of bases.

4. A clear understanding of the relationship that exists between the attack and the defense.

5. The development of mobile operations.

6. Correct command.

The enemy, though numerically weak, is strong in the quality of his troops and their equipment; we, on the other hand, are strong numerically but weak as to quality. These considerations have been taken into account in the development of the policy of tactical offense, tactical speed, and tactical operations on exterior lines in a war that, strategically speaking, is defensive in character, protracted in nature, and conducted along interior lines. Our strategy is based on these conceptions. They must be kept in mind in the conduct of all operations.

Although the element of surprise is not absent in orthodox warfare, there are fewer opportunities to apply it than there are during guerrilla hostilities. In the latter, speed is essential. The movements of guerrilla troops must be secret and of supernatural rapidity; the enemy must be taken unaware, and the action entered speedily. There can be no procrastination in the execution of plans; no assumption of a negative or passive defense; no great dispersion of forces in many local engagements. The basic method is the attack in a violent and deceptive form.

While there may be cases where the attack will extend over a period of several days (if that length of time is necessary to annihilate an enemy group), it is more profitable to launch and push an attack with maximum speed. The tactics of defense have no place in the realm of guerrilla warfare. If a delaying action is necessary, such places as defiles, river crossings, and villages offer the most suitable conditions, for it is in such places that the enemy's arrangements may be disrupted and he may be annihilated.

The enemy is much stronger than we are, and it is true

that we can hinder, distract, disperse, and destroy him only if we disperse our own forces. Although guerrilla warfare is the warfare of such dispersed units, it is sometimes desirable to concentrate in order to destroy an enemy. Thus, the principle of concentration of force against a relatively weaker enemy is applicable to guerrilla warfare.

We can prolong this struggle and make of it a protracted war only by gaining positive and lightning-like tactical decisions; by employing our manpower in proper concentrations and dispersions; and by operating on exterior lines in order to surround and destroy our enemy. If we cannot surround whole armies, we can at least partially destroy them; if we cannot kill the Japanese, we can capture them. The total effect of many local successes will be to change the relative strengths of the opposing forces. The destruction of Japan's military power, combined with the international sympathy for China's cause and the revolutionary tendencies evident in Japan, will be sufficient to destroy Japanese imperialism.

We will next discuss initiative, alertness, and the matter of careful planning. What is meant by initiative in warfare? In all battles and wars, a struggle to gain and retain the initiative goes on between the opposing sides, for it is the side that holds the initiative that has liberty of action. When an army loses the initiative, it loses its liberty; its role becomes passive; it faces the danger of defeat and destruction.

It is more difficult to obtain the initiative when defending on interior lines than it is while attacking on exterior lines. This is what Japan is doing. There are, however, several weak points as far as Japan is concerned. One of these is lack of sufficient manpower for the task; another is her cruelty to the inhabitants of conquered areas; a third is the underestimation of Chinese strength, which has resulted in the differences between military cliques, which, in turn, have been productive of many mistakes in the direction of her military forces. For instance, she has been gradually compelled to increase her manpower in China while, at the same time, the many arguments over plans of operations and disposition of troops have resulted in the loss of good opportunities for improvement of her strategical position. This explains the fact that although the Japanese are frequently able to surround large bodies of Chinese troops, they have never yet been able to capture more than a few. The Japanese military machine is thus being weakened by insufficiency of manpower, inadequacy of resources, the barbarism of her troops, and the general stupidity that has characterized the conduct of operations. Her offensive continues unabated, but because of the weaknesses pointed out, her attack must be limited in extent. She can never conquer China. The day will come—indeed, already has in some areas—when she will be forced into a passive role. When hostilities commenced, China was passive, but as we enter the second phase of the war, we find ourselves pursuing a strategy of mobile warfare, with both guerrillas and regulars operating on exterior lines. Thus, with each passing day, we seize some degree of initiative from the Japanese.

The matter of initiative is especially serious for guerrilla forces, who must face critical situations unknown to regular troops. The superiority of the enemy and the lack of unity and experience within our own ranks may be cited. Guerrillas can, however, gain the initiative if they keep in mind the weak points of the enemy. Because of the enemy's insufficient manpower, guerrillas can operate over vast territories; because he is a foreigner and a barbarian, guerrillas can gain the confidence of millions of their countrymen; because of the stupidity of enemy commanders, guerrillas can make full use of their own cleverness. Both guerrillas and regulars must exploit these enemy weaknesses while, at the same time, our own are remedied. Some of our weaknesses are apparent only and are, in actuality, sources of strength. For example, the very fact that most guerrilla groups are small makes it desirable and advantageous for them to appear and disappear in the enemy's rear. With such activities, the enemy is simply unable to cope. A similar liberty of action can rarely be obtained by orthodox forces.

When the enemy attacks the guerrillas with more than one column, it is difficult for the latter to retain the initiative. Any error, no matter how slight, in the estimation of the situation is likely to result in forcing the guerrillas into a passive role. They will then find themselves unable to beat off the attacks of the enemy.

It is apparent that we can gain and retain the initiative only by a correct estimation of the situation and a proper arrangement of all military and political factors. A too pessimistic estimate will operate to force us into a passive position, with consequent loss of initiative; an overly optimistic estimate, with its rash ordering of factors, will produce the same result.

No military leader is endowed by heaven with an ability to seize the initiative. It is the intelligent leader who does so after a careful study and estimate of the situation and arrangement of the military and political factors involved. When a guerrilla unit, through either a poor estimate on the part of its leader or pressure from the enemy, is forced into a passive position, its first duty is to extricate itself. No method can be prescribed for this, as the method to be employed will, in every case, depend on the situation. One can, if necessary, run away. But there are times when the situation seems hopeless and, in reality, is not so at all. It is at such times that the good leader recognizes and seizes the moment when he can regain the lost initiative.

Let us revert to alertness. To conduct one's troops with alertness is an essential of guerrilla command. Leaders must realize that to operate alertly is the most important factor in gaining the initiative and vital in its effect on the relative situation that exists between our forces and those of the enemy. Guerrilla commanders adjust their operations to the enemy situation, to the terrain, and to prevailing local conditions. Leaders must be alert to sense changes in these factors and make necessary modifications in troop dispositions to accord with them. The leader must be like the fisherman, who, with his nets, is able both to cast them and to pull them out in awareness of the depth of the water, the strength of the current, or the presence of any obstructions that may foul them. As the fisherman controls his nets through the lead ropes, so the guerrilla leader maintains contact with and control over his units. As the fisherman must change his position, so must the guerrilla commander. Dispersion, concentration, constant change of position—it is in these ways that guerrillas employ their strength.

In general, guerrilla units disperse to operate:

1. When the enemy is in overextended defense, and sufficient force cannot be concentrated against him, guerrillas must disperse, harass him, and demoralize him.

2. When encircled by the enemy, guerrillas disperse to withdraw.

3. When the nature of the ground limits action, guerrillas disperse.

4. When the availability of supplies limits action, they disperse.

5. Guerrillas disperse in order to promote mass movements over a wide area.

Regardless of the circumstances that prevail at the time of dispersal, caution must be exercised in certain matters:

1. A relatively large group should be retained as a central force. The remainder of the troops should not be divided into groups of absolutely equal size. In this way, the leader is in a position to deal with any circumstances that may arise.

2. Each dispersed unit should have clear and definite responsibilities. Orders should specify a place to which to

proceed, the time of proceeding, and the place, time, and method of assembly.

Guerrillas concentrate when the enemy is advancing upon them, and there is opportunity to fall upon him and destroy him. Concentration may be desirable when the enemy is on the defensive and guerrillas wish to destroy isolated detachments in particular localities. By the term "concentrate," we do not mean the assembly of all man-power but rather of only that necessary for the task. The remaining guerrillas are assigned missions of hindering and delaying the enemy, of destroying isolated groups, or of conducting mass propaganda.

In addition to the dispersion and concentration of forces, the leader must understand what is termed "alert shifting." When the enemy feels the danger of guerrillas, he will generally send troops out to attack them. The guerrillas must consider the situation and decide at what time and at what place they wish to fight. If they find that they cannot fight, they must immediately shift. Then the enemy may be destroyed piecemeal. For example, after a guerrilla group has destroyed an enemy detachment at one place, it may be shifted to another area to attack and destroy a second detachment. Sometimes, it will not be profitable for a unit to become engaged in a certain area, and in that case, it must move immediately.

When the situation is serious, the guerrillas must move with the fluidity of water and the ease of the blowing wind. Their tactics must deceive, tempt, and confuse the enemy. They must lead the enemy to believe that they will attack him from the east and north, and they must then strike him from the west and the south. They must strike, then rapidly disperse. They must move at night.

Guerrilla initiative is expressed in dispersion, concentration, and the alert shifting of forces. If guerrillas are stupid and obstinate, they will be led to passive positions and severely damaged. Skill in conducting guerrilla operations, however, lies not in merely understanding the things we have discussed but rather in their actual application on the field of battle. The quick intelligence that constantly watches the ever-changing situation and is able to seize on the right moment for decisive action is found only in keen and thoughtful observers.

Careful planning is necessary if victory is to be won in guerrilla war, and those who fight without method do not understand the nature of guerrilla action. A plan is necessary regardless of the size of the unit involved; a prudent plan is as necessary in the case of the squad as in the case of the regiment. The situation must be carefully studied, then an assignment of duties made. Plans must include both political and military instruction; the matter of supply and equipment, and the matter of cooperation with local civilians. Without study of these factors, it is impossible either to seize the initiative or to operate alertly. It is true that guerrillas can make only limited plans, but even so, the factors we have mentioned must be considered.

The initiative can be secured and retained only following a positive victory that results from attack. The attack must be made on guerrilla initiative; that is, guerrillas must not permit themselves to be maneuvered into a position where they are robbed of initiative and where the decision to attack is forced upon them. Any victory will result from careful planning and alert control. Even in defense, all our efforts must be directed toward a resumption of the attack, for it is only by attack that we can extinguish our enemies and preserve ourselves. A defense or a withdrawal is entirely useless as far as extinguishing our enemies is concerned and of only temporary value as far as the conservation of our forces is concerned. This principle is valid both for guerrillas and regular troops. The differences are of degree only; that is to say, in the manner of execution.

The relationship that exists between guerrillas and the orthodox forces is important and must be appreciated. Generally speaking, there are three types of cooperation between guerrillas and orthodox groups. These are:

1. Strategical cooperation.
2. Tactical cooperation.
3. Battle cooperation.

Guerrillas who harass the enemy's rear installations and hinder his transport are weakening him and encouraging the national spirit of resistance. They are cooperating strategically. For example, the guerrillas in Manchuria had no functions of strategical cooperation with orthodox forces until the war in China started. Since that time, their function of strategical cooperation is evident, for if they can kill one enemy, make the enemy expend one round of ammunition, or hinder one enemy group in its advance southward, our powers of resistance here are proportionately increased. Such guerrilla action has a positive action on the enemy nation and on its troops, while, at the same time, it encourages our own countrymen. Another example of strategical cooperation is furnished by the guerrillas who operate along the P'ing-Sui, P'ing-Han, Chin-P'u, T'ung-Pu, and Cheng-T'ai railways. This cooperation began when the invader attacked, continued during the period when he held garrisoned cities in the areas, and was intensified when our regular forces counterattacked, in an effort to restore the lost territories.

As an example of tactical cooperation, we may cite the operations at Hsing-K'ou, when guerrillas both north and south of Yeh Men destroyed the T'ung-P'u railway and the motor roads near P'ing Hsing Pass and Yang Fang K'ou. A number of small operating bases were established, and organized guerrilla action in Shansi complemented the activities of the regular forces both there and in the defense of Honan. Similarly, during the south Shantung campaign, guerrillas in the five northern provinces cooperated with the army's operation on the Hsuchow front.

Guerrilla commanders in rear areas and those in command of regiments assigned to operate with orthodox units must cooperate in accordance with the situation. It is their function to determine weak points in the enemy dispositions, to harass them, to disrupt their transport, and to undermine their morale. If guerrilla action were independent, the results to be obtained from tactical cooperation would be lost and those that result from strategical cooperation greatly diminished. In order to accomplish their mission and improve the degree of cooperation, guerrilla units must be equipped with some means of rapid communication. For this purpose, two-way radio sets are recommended.

Guerrilla forces in the immediate battle area are responsible for close cooperation with regular forces. Their principal functions are to hinder enemy transport, to gather information, and to act as outposts and sentinels. Even without precise instructions from the commander of the regular forces, these missions, as well as any others that contribute to the general success, should be assumed.

The problem of establishment of bases is of particular importance. This is so because this war is a cruel and protracted struggle. The lost territories can be restored only by a strategical counterattack, and this we cannot carry out until the enemy is well into China. Consequently, some part of our country—or, indeed, most of it—may be captured by the enemy and become his rear area. It is our task to develop intensive guerrilla warfare over this vast area and convert the enemy's rear into an additional front. Thus the enemy will never be able to stop fighting. In order to subdue the occupied territory, the enemy will have to become increasingly severe and oppressive.

A guerrilla base may be defined as an area, strategically located, in which the guerrillas can carry out their duties of training, self-preservation and development. Ability to fight a war without a rear area is a fundamental characteristic of guerrilla action, but this does not mean that guerrillas can exist and function over a long period of time without the development of base areas. History shows us many examples of peasant revolts that were unsuccessful, and it is fanciful to believe that such movements, characterized by banditry and brigandage, could succeed in this era of improved communications and military equipment. Some guerrilla leaders seem to think that those qualities are present in today's movement, and before such leaders can comprehend the importance of base areas in the long-term war, their minds must be disabused of this idea.

The subject of bases may be better understood if we consider:

1. The various categories of bases.
2. Guerrilla areas and base areas.
3. The establishment of bases.
4. The development of bases.

Guerrilla bases may be classified according to their location as: first, mountain bases; second, plains bases; and, last, river, lake, and bay bases. The advantages of bases in mountainous areas are evident. Those which are now established are at Ch'ang P'o Chan, Wu Tai Shan, Taiheng Shan, Tai Shan, Yen Shan, and Mao Shan. These bases are strongly protected. Similar bases should be established in all enemy rear areas.

Plains country is generally not satisfactory for guerrilla operating bases, but this does not mean that guerrilla warfare cannot flourish in such country or that bases cannot be established there. The extent of guerrilla development in Hopeh and west Shantung proves the opposite to be the case. Whether we can count on the use of these bases over long periods of time is questionable. We can, however, establish small bases of a seasonal or temporary nature. This we can do because our barbaric enemy simply does not have the manpower to occupy all the areas he has over-run and because the population of China is so numerous that a base can be established anywhere. Seasonal bases in plains country may be established in the winter when the rivers are frozen over, and in the summer when the crops are growing. Temporary bases may be established when the enemy is otherwise occupied. When the enemy advances, the guerrillas who have established bases in the plains area are the first to engage him. Upon their with-drawal into mountainous country, they should leave be-hind them guerrilla groups dispersed over the entire area. Guerrillas shift from base to base on the theory that they must be one place one day and another place the next.

There are many historical examples of the establishment of bases in river, bay, and lake country, and this is one aspect of our activity that has so far received little atten-tion. Red guerrillas held out for many years in the Hungtze Lake region. We should establish bases in the Hungtze and Tai areas and along rivers and watercourses in territory controlled by the enemy so as to deny him access to, and free use of, the water routes.

There is a difference between the terms base area and guerrilla area. An area completely surrounded by territory occupied by the enemy is a "base area." Wu Tai Shan, Tai Shan, and Taiheng Shan are examples of base areas. On the other hand, the area east and north of Wu Tai Shan (the Shansi-Hopeh-Chahar border zone) is a guerrilla area. Such areas can be controlled by guerrillas only while they actually physically occupy them. Upon their departure, control reverts to a puppet pro-Japanese government. East Hopeh, for example, was at first a guerrilla area rather than a base area. A puppet government functioned there. Eventually, the people, organized and inspired by guerrillas from the Wu Tai mountains, assisted in the transformation of this guerrilla area into a real base area. Such a task is extremely difficult, for it is largely dependent upon the degree to which the people can be inspired. In certain garrisoned areas, such as the cities and zones contiguous to the railroads, the guerrillas are unable to drive the Japanese and puppets out. These areas remain guerrilla areas. At other times, base areas might become guerrilla areas due either to our own mistakes or to the activities of the enemy.

Obviously, in any given area in the war zone, any one of three situations may develop: The area may remain in Chinese hands; it may be lost to the Japanese and puppets; or it may be divided between the combatants. Guerrilla leaders should endeavor to see that either the first or the last of these situations is assured.

Another point essential in the establishment of bases is the cooperation that must exist between the armed guerrilla bands and the people. All our strength must be used to spread the doctrine of armed resistance to Japan, to arm the people, to organize self-defense units, and to train guerrilla bands. This doctrine must be spread among the people, who must be organized into anti-Japanese groups. Their political instincts must be sharpened and their martial ardor increased. If the workers, the farmers, the lovers of liberty, the young men, the women, and the children are not organized, they will never realize their own anti-Japanese power. Only the united strength of the people can eliminate traitors, recover the measure of political power that has been lost, and conserve and improve what we still retain.

We have already touched on geographic factors in our discussion of bases, and we must also mention the economic aspects of the problem. What economic policy should be adopted? Any such policy must offer reasonable protection to commerce and business. We interpret "reasonable protection" to mean that people must contribute money in proportion to the money they have. Farmers will be required to furnish a certain share of their crops to guerrilla troops. Confiscation, except in the case of businesses run by traitors, is prohibited.

Our activities must be extended over the entire periphery of the base area if we wish to attack the enemy's bases and thus strengthen and develop our own. This will afford us opportunity to organize, equip, and train the people, thus furthering guerrilla policy as well as the national policy of protracted war. At times, we must emphasize the development and extension of base areas; at other times, the organization, training, or equipment of the people.

Each guerrilla base will have its own peculiar problems of attack and defense. In general, the enemy, in an endeavor to consolidate his gains, will attempt to extinguish guerrilla bases by dispatching numerous bodies of troops over a number of different routes. This must be anticipated and the encirclement broken by counterattack. As such enemy columns are without reserves, we should plan on using our main forces to attack one of them by surprise and devote our secondary effort to continual hindrance and harassment. At the same time, other forces should isolate enemy garrison troops and operate on their lines of supply and communication. When one column has been disposed of, we may turn our attention to one of the others. In a base area as large as Wu Tai Shan, for example, there are four or five military subdivisions. Guerrillas in these subdivisions must cooperate to form a primary force to counterattack the enemy, or the area from which he came, while a secondary force harasses and hinders him.

After defeating the enemy in any area, we must take advantage of the period he requires for reorganization to press home our attacks. We must not attack an objective we are not certain of winning. We must confine our operations to relatively small areas and destroy the enemy and traitors in those places.

When the inhabitants have been inspired, new volunteers accepted, trained, equipped, and organized, our operations may be extended to include cities and lines of communication not strongly held. We may hold these at least for temporary (if not for permanent) periods. All these are our duties in offensive strategy. Their object is to lengthen the period that the enemy must remain on the defensive. Then our military activities and our organization work among the masses of the people must be zealously expanded; and with equal zeal, the strength of the enemy attacked and diminished. It is of great importance that guerrilla units be rested and instructed. During such times as the enemy is on the defensive, the troops may get some rest and instruction may be carried out.

The development of mobile warfare is not only possible but essential. This is the case because our current war is a desperate and protracted struggle. If China were able to conquer the Japanese bandits speedily and to recover her lost territories, there would be no question of long-term war on a national scale. Hence, there would be no question of the relation of guerrilla warfare and the war of movement. Exactly the opposite is actually the case. In order to ensure the development of guerrilla hostilities into mobile warfare of an orthodox nature, both the quantity and quality of guerrilla troops must be improved. Primarily, more men must join the armies; then the quality of equipment and standards of training must be improved. Political training must be emphasized and our organization, the technique of handling our weapons, our tactics—all must be improved. Our internal discipline must be strengthened. The soldiers must be educated politically. There must be a gradual change from guerrilla formations to orthodox regimental organization. The necessary bureaus and staffs, both political and military, must be provided. At the same time, attention must be paid to the creation of suitable supply, medical, and hygiene units. The standards of equipment must be raised and types of weapons increased. Communication equipment must not be forgotten. Orthodox standards of discipline must be established.

Because guerrilla formations act independently and because they are the most elementary of armed formations, command cannot be too highly centralized. If it were, guerrilla action would be too limited in scope. At the same time, guerrilla activities, to be most effective, must be coordinated, not only insofar as they themselves are concerned, but additionally with regular troops operating in the same areas. This coordination is a function of the war-zone commander and his staff.

In guerrilla base areas, the command must be centralized for strategical purposes and decentralized for tactical purposes. Centralized strategical command takes care of the general management of all guerrilla units, their coordination within war zones, and the general policy regarding guerrilla base areas. Beyond this, centralization of command will result in interference with subordinate units, as, naturally, the tactics to apply to concrete situations can be determined only as these various situations arise. This is true in orthodox warfare when communications between lower and higher echelons break down. In a word, proper guerrilla policy will provide for unified strategy and independent activity.

Each guerrilla area is divided into districts and these in turn are divided into subdistricts. Each subdivision has its appointed commander, and while general plans are made by higher commanders, the nature of actions is determined by inferior commanders. The former may suggest the nature of the action to be taken but cannot define it. Thus inferior groups have more or less complete local control.

APPENDIX

TABLE 1

Organization of an Independent Guerrilla Company

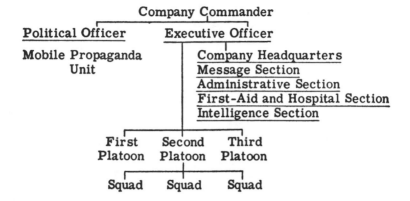

Table of Organization, Guerrilla Company

Rank	Personnel	Arm
Company Leader	1	Pistol
Political Officer	1	Pistol
Executive Officer	1	Pistol
Company Headquarters		
Message Section Chief	1	
Signal	1	
Administrative Section Chief	1	Rifle
Public Relations	3	Rifle
Duty Personnel	2	
Barber	1	
Cooks	10	
Medical Section Chief	1	
Assistant	1	
First Aid and Nursing	4	
Intelligence Section Chief	1	Rifle
Intelligence	9	Rifle
Platoon Leaders	3	Rifle
Squad Leaders	9	Rifle
Nine Squads (8 each)	72	Rifle
TOTAL	122	3 Pistols 98 Rifles

NOTES

1. Each squad consists of from 9 to 11 men. In case men or arms are not sufficient, the third platoon may be dispensed with or one squad organized as company headquarters.

2. The mobile propaganda unit consists of members of the company who are not relieved of primary duties except to carry out propaganda when they are not fighting.

3. If there is insufficient personnel, the medical section is not separately organized. If there are only two or three medical personnel, they may be attached to the administrative section.

4. If there is no barber, it is unimportant. If there is an insufficient number of cooks, any member of the company may be designated to prepare food.

5. Each combatant soldier should be armed with the rifle. If there are not enough rifles, each squad should have two or three. Shotguns, lances, and big swords can also be furnished. The distribution of rifles does not have to be equalized in platoons. As different missions are assigned to platoons, it may be necessary to give one platoon more rifles than the others.

6. The strength of a company should at the most be 180, divided into 12 squads of 11 men each. The minimum strength of a company should be 82 men, divided into 6 squads of 9 men each.

TABLE 2

ORGANIZATION OF AN INDEPENDENT GUERRILLA BATTALION

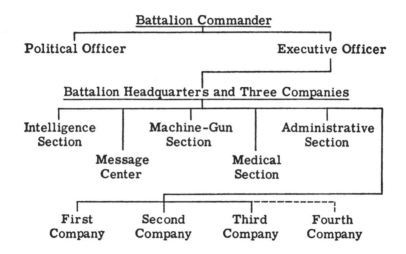

NOTES

1. Total headquarters of an independent guerrilla battalion may vary from a minimum of 46 to a maximum of 110.

2. When there are 4 companies to a battalion, regimental organization should be used.

3. Machine-gun squads may be heavy or light. A light machine-gun squad has from 5 to 7 men. A heavy machine-gun squad has from 7 to 9 men.

4. The intelligence section is organized in from 2 to 4 squads, at least one of which is made up of plain-clothes men. If horses are available, one squad should be mounted.

5. If no men are available for stretcher-bearers, omit them and use the cooks or ask aid from the people.

6. Each company must have at least 25 rifles. The remaining weapons may be bird guns, big swords, or locally made shotguns.

TABLE OF ORGANIZATION, GUERRILLA BATTALION (INDEPENDENT)

RANK	PERSONNEL	ARM
Battalion Commander	1	Pistol
Political Officer	1	Pistol
Executive Officer	1	Pistol
Battalion Headquarters		
Signal Section	2	
Administrative Section		
Section Chief	1	Carbine
Runner	1	Carbine
Public Relations	10	Carbine
Duty Personnel	2	
Barbers	3	
Supply	1	
Cooks	10	
Medical Section		
Medical Officer	1	
Stretcher-Bearers	6	
Nursing	4	
Intelligence Section		
Section Chief	1	Pistol
Intelligence	30	Pistol
Machine-Gun Section	As Available	As Available
Total, Headquarters	75	34 Pistols 12 Carbines
Three Companies (122 each)	366	9 Pistols 288 Carbines
TOTAL	441	43 Pistols 300 Rifles

TABLE 3

ORGANIZATION OF AN INDEPENDENT GUERRILLA REGIMENT

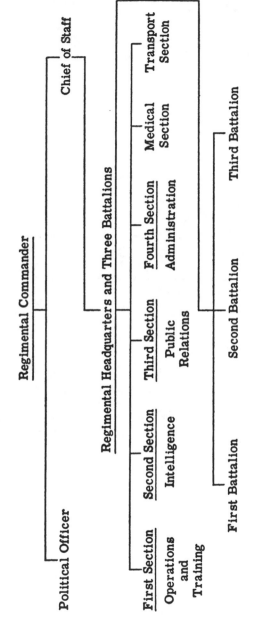

NOTES

1. See Tables 1 and 2 for company and battalion organization.
2. Battalions and companies have no transport sections.
3. The hand weapon may be either revolver or pistol. Of these, each battalion should have more than 100.

TABLE OF ORGANIZATION, GUERRILLA REGIMENT

RANK	PERSONNEL	ARM
Regimental Commander	1	Pistol
Political Officer	1	Pistol
Chief of Staff	1	Pistol
Operations Section		
Operations Officer	1	Pistol
Clerks	15	
Intelligence Section		
Intelligence Officer	1	Pistol
Personnel	36	Pistols
Public-Relations Section		
Public-Relations Officer	1	Pistol
Personnel	36	Carbines
Administrative Section		
Administrative Officer	1	Pistol
Clerks	15	Pistol
Runner	1	
Transport Section		
Chief of Section	1	Pistol
Finance	1	
Traffic Manager	1	Pistol
Supply	1	
Drivers	5	
Medical Section		
Chief of Section	1	
Doctors	2	
Nurses	15	
Total, Regimental Headquarters	137	60 Pistols 36 Carbines
Three Battalions (441 each)	1323	124 Pistols 900 Rifles
TOTAL	1460	184 Pistols 936 Rifles

TABLE 4

ORGANIZATION OF INDEPENDENT GUERRILLA BRIGADE (OR DIVISION)

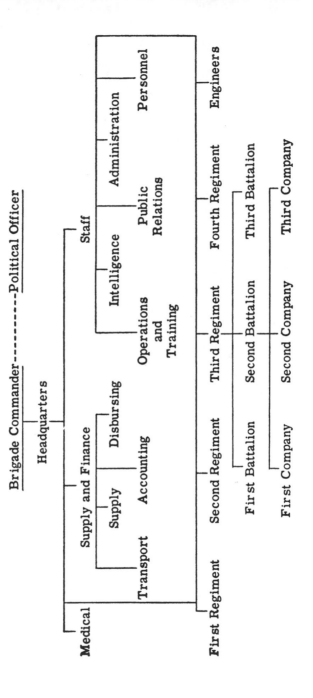